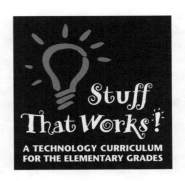

Stuff
That Works!
A TECHNOLOGY CURRICULUM
FOR THE ELEMENTARY GRADES

Packaging & Other Structures

Gary Benenson and James L. Neujahr
Project Directors, City Technology

Heinemann
Portsmouth, NH

Heinemann
A division of Reed Elsevier Inc.
361 Hanover Street
Portsmouth, NH 03801–3912
www.heinemann.com

Offices and agents throughout the world

Stuff That Works!
City College of New York

140 Street & Convent Avenue, Room T233
New York, New York 10031
(212) 650-8389 tel.; (212) 650-8013 fax
citytechnology@ccny.cuny.edu

This project was supported, in part, by the
National Science Foundation
Opinions expressed are those of the authors and not necessarily those of the Foundation

Project Staff
Gary Benenson, *Project Director, City College of New York, School of Engineering*
James L. Neujahr, *Project Co-Director, City College of New York, School of Education*
Dorothy Bennett, *Education Development Center/Center for Children and Technology*
Terri Meade, *Education Development Center/Center for Children and Technology*

Advisory Board
William Barowy, *Lesley College*
David Chapin, *City University of New York*
Alan Feigenberg, *City College of New York*
Ed Goldman, *Brooklyn Technical High School*

Patricia Hutchinson, *The College of New Jersey*
Neville Parker, *City College of New York*
Peter Sellwood, *Consultant, United Kingdom*
Ron Todd, *The College of New Jersey*

Teacher Associates/Coauthors
Katherine Aguiar, *CES 42, Bronx, NY*
Helen deCandido, *Retired*
Mary Flores, *CES 42, Bronx, NY*
Angel Gonzalez, *Family Academy, New York, NY*
Michael Gatton, *IS 143M, New York, NY*
Theresa Luongo, *Central Park East #2, New York, NY*
Roslyn Odinga, *CES 126, Bronx, NY*

Felice Piggott, *PS 145, New York, NY*
Annette Purnell, *CES 42, Bronx, NY*
Minerva Rivera, *Harbor Academy, New York, NY*
Sandra Skea, *Mott Hall School, New York, NY*
Christine Smith, *IS 164, New York, NY*
Verona Williams, *PS 60, Bronx, NY*

Production Staff
Gary Benenson, *General Editor and Lead Author*
Lorin Driggs, *Editor*
Doris Halle Design NYC, *Design and Graphics*
Maria Politarhos, *Photography*
Juana Maria Page, *Illustrations*

Library of Congress Cataloging-in-Publication Data
Benenson, Gary.
    Stuff that works! : a technology curriculum for the elementary grades / Gary Benenson and James L. Neujahr.
        p. cm.
    Includes bibliographical references.
    Contents: [v. 2] Packaging & other structures
    ISBN 0-325-00469-2
    1. Technology—Study and Teaching(Elementary)—United States. I. Neujahr, James L., 1939– . II. Title.

T72 .B46 2002
372.3'5—dc21   2001059398

Printed in the United States of America on acid-free paper

06 05 04 03   VP   2 3 4 5

CONTENTS

FOREWORD

IN A WORLD INCREASINGLY DEPENDENT ON TECHNOLOGY—where new ideas and tools pervade our personal and civic lives and where important choices hinge on our knowledge of how things and people work—the imperative that all students should learn to understand and use technology well should be obvious. Yet in the American curriculum, still overstuffed with tradition and trivia, there is little room in the day for learning and teaching about important ideas from technology and very few resources for educators who want to engage their students in learning for the 21st century.

Stuff That Works! is a groundbreaking curriculum. It provides a set of carefully chosen and designed activities that will engage elementary students with the core ideas and processes of technology (or engineering, if you prefer). Elementary school is the ideal place to begin learning about technology. It is a time in students' development when they are ready and eager to take on concrete rather than abstract ideas. The concepts and skills presented in *Stuff That Works!* will support more advanced learning in mathematics, science, and technology as students move up through the grades.

But there is much more to *Stuff That Works!* than a set of activities. As a matter of fact, the activities make up less than a third of the pages. *Stuff That Works!* also includes helpful resources for the teacher such as clear discussions of the important ideas and skills from technology that their students should be learning; stories of how the materials have been used in real classrooms; suggestions for outside reading; guidance for assessing how well their students are doing; and tips on implementation. I hope teachers will take time to make full use of these valuable resources as they use *Stuff That Works!* If they do, they can help their students take the first, critical steps towards technological literacy and success in and beyond school.

George D. Nelson, Director
*American Association for
the Advancement of Science (AAAS)
Project 2061*

INTRODUCTION

What Is Technology?

Stuff That Works! *Packaging and Other Structures* will introduce you to a novel and very engaging approach to the study of technology at the elementary school level. In education today, the word *technology* is most often associated with learning how to use computers, and that is certainly important. But learning how to use a particular kind of technology is not the same thing as learning how and why the technology works. Children learn about computers as *users* rather than as students of how computers work or of how to design them. In fact, computer analysis and design require technical knowledge that is beyond most adults, let alone elementary-aged children. Fortunately, there are many other examples of technology that are much more accessible than computers and that present many of the same issues as computers and other "high-tech" devices.

The purpose of technology is to solve practical problems by means of devices, systems, procedures, and environments that improve people's lives in one way or another. Understood this way, a computer is no more an example of technology than...

- the cardboard box it was shipped in,
- the arrangement of the computer and its peripherals on the table,
- the symbol next to the printer's ON/OFF switch,
- or the ballpoint pen the printer replaces as a writing device.

A box, a plan for the use of table space, an ON/OFF symbol, and a pen are examples of technologies you and your students will explore in this and the other *Stuff That Works!* guides.

The *Stuff That Works!* approach is based on artifacts and systems that are all around us and available for free or at very low cost. You need not be a technical guru or rich in resources to engage yourself and your students in technology. The materials needed for *Packaging and Other Structures* are nearly all discarded items such as empty bottles, boxes, and bags; cushioning materials such as Styrofoam and bubble wrap, plus a few common school supplies such as blocks, tape, and glue.

Why Study Technology in Elementary School?

Below is a graphic summary of the process of "doing" technology as we present it in this book. The study of technology challenges students to identify and solve problems, build understanding, develop and apply competence and knowledge in a variety of processes and content areas, including science, mathematics, language arts, and social interaction.

The teachers who field-tested these materials underscored that these activities helped their students to:

- observe and describe phenomena in detail;
- explore real objects and situations by creating models and other representations;
- identify salient aspects of problems;
- solve authentic problems;
- use evidence-based reasoning;
- apply the scientific method;
- ask thoughtful questions (beyond the yes or no variety);
- communicate in oral, written, and graphic form;
- collaborate effectively with others.

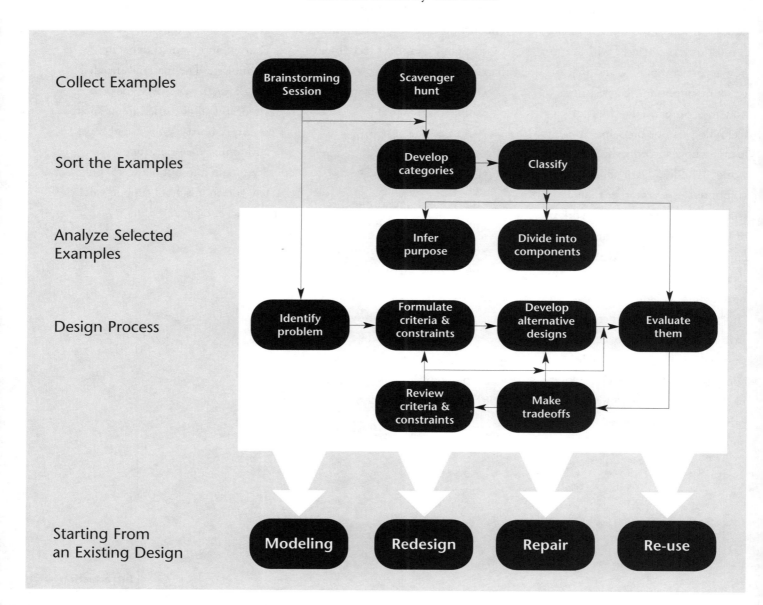

Educational Goals for Packaging and Other Structures

Packaging and Other Structures explores how bags, boxes, cartons, and bottles work to contain, protect, dispense, and display products. All kinds of packaging materials are examples of structures, which are technologies designed to support mechanical loads. The content and activities in this book will help you to meet the following educational goals:

- Develop fundamental themes of systems, material properties, spatial relationships, and trade-offs;
- Motivate and illustrate concepts of force, structure, load and failure; compression, tension, and shear; repair, redesign, and re-use;
- Demystify common artifacts, and by extension, technology in general;
- Develop process skills in observation, classification, generalization, prediction, control of variables, design, and evaluation;
- Provide rich opportunities for group work;
- Develop environmental awareness.

How This Guide Is Organized

Each **Stuff That Works!** guide is organized into the following chapters.

Chapter 1. *Appetizers* suggests some things you can do for yourself, to become familiar with the topic. You can do these activities at home, using only found materials. They will help you to recognize some of the technology that is all around you, and offer ways of making sense of it.

Chapter 2. *Concepts* develops the main ideas that can be taught for and through the topic. These include ideas from science, math, social studies, and art, as well as technology. It also reviews what is known from relevant cognitive research.

Chapter 3. *Activities* contains a variety of classroom projects and units related to the topic, including those referred to in Chapter 4. Each activity includes prerequisites, goals, skills and concepts; materials, references to standards and teacher tips; and sample worksheets.

Chapter 4. *Stories* presents teachers' narratives about what happened in their own classrooms. Their accounts include photos, samples of children's work and children's dialog. Commentary by project staff connects the teachers' accounts with the concepts developed in Chapter 2.

Chapter 5. *Resources* provides a framework supporting the implementation of the activities. It includes an annotated bibliography of children's literature and a discussion of assessment principles and opportunities.

Chapter 6. *About Standards* shows how the activities and ideas in this book address national standards in technology, science, math, and English language arts (ELA).

How to Use This Guide

Different teachers will obviously come to this book with different needs and objectives. However, regardless of your background, instructional approach, and curricular goals, *we strongly recommend that you begin with Chapter 1, "Appetizers."* There is simply no better way to become acquainted with a topic and to understand what your students will be facing than to try out some of the ideas and activities for yourself. Chapter 1 guides you through that process.

The content and approach presented in *Packaging and Other Structures* are based on the premise that processes of design are central to the practice of technology, just as inquiry is the central activity of science. While no two design problems are the same, there are some features that characterize any design task:

- It should solve a problem of some sort.
- It must have more than one possible solution.
- There must be an effort to test the design.

A problem is like a trigger that initiates a design process. Often the problem is not well-formulated, a vague kind of "wouldn't it be nice if …" In making the problem more specific, it is often helpful to list some criteria the design must address. In trying to satisfy these criteria, the designer is never completely free to do whatever he or she wants. There are always constraints, which could involve cost, safety, ease of use, and a host of other considerations.

Packaging and Other Structures presents a number of activities that include elements of design, but are not full-scale design projects. These elements of design are modeling, redesign, repair, and re-use.

- **Modeling** requires both a very close look at the original design and its modification to incorporate the use of different materials. A lot can be learned by observing how the substitution of materials affects the operation of the model. Modeling is only one kind of design activity that starts with an existing solution.

- **Redesign** starts with an existing but inadequate design. It involves analyzing the weaknesses of the original design and then figuring out how to correct them.

- **Repair** is a variant of redesign. It takes place after the existing design has already failed. Redesign and repair projects often use new materials or techniques to accomplish the original purpose.

- **Re-use** is a complementary kind of design activity where the original materials are used for a new purpose.

The concepts of redesign, repair, and re-use are of particular importance in a society that has been widely criticized for its wasteful practices. These three concepts are considerably more accessible than the more widely advanced notion of recycling, whose full implementation requires expensive equipment and specialized technical knowledge.

There is no one way to do design. It is a non-linear, messy process that typically begins with very incomplete information. Additional criteria become apparent as the design is implemented and tested. New constraints appear that were not originally evident. It is often necessary to backtrack and revise the original specifications. Such a messy process may seem contrary to the work you usually expect to see happening in your classroom. However, we encourage you to embrace the messiness! It will justify itself by improving students' competence in reasoning, problem-solving, and ability to communicate not only what they are doing but also why they are doing it and what results they expect.

A Brief History of *Stuff That Works!*

The guides in the *Stuff That Works!* series were developed through collaboration among three different kinds of educators:

- Two college professors, one from the School of Education of City College of the City University of New York, and the other from the City College School of Engineering;

- Two educational researchers from the Center for Children and Technology of the Education Development Center (CCT/ EDC);

- Thirty New York City elementary educators who work in the South Bronx, Harlem, and Washington Heights.

This last group included science specialists, early childhood educators, special education teachers, a math specialist, a language arts specialist, and regular classroom teachers from grades two through seven. In teaching experience, they ranged from first-year teachers to veterans with more than 20 years in the classroom.

During the 1997-98 and 1998-99 academic years, the teachers participated in workshops that engaged them in sample activities and also provided opportunities for sharing and discussion of classroom experiences. The workshop activities then became the basis for classroom implementation.

The teachers were encouraged to modify the workshop activities and extend them in accordance with their own teaching situations, their ideas, and their children's interests.

The teachers, project staff, and the research team collaborated to develop a format for documenting classroom outcomes in the form of portfolios. These portfolios included the following items:

- lesson worksheets describing the activities and units implemented in the classroom, including materials used, teacher tips and strategies, and assessment methods;

- narrative descriptions of what actually happened in the classroom;

- samples of students' work, including writing, maps and drawings, and dialogue; and

- the teachers' own reflections on the activities.

The lesson worksheets became the basis for the **Activities** (Chapter 3) of each guide. The narratives, samples of student work, and teacher reflections formed the core of the **Stories** (Chapter 4). At the end of the two years of curriculum development and pilot testing, the project produced five guides in draft form.

During the 1999-2000 academic year, the five draft guides were field-tested at five sites, including two in New York City, one suburban New York site, and one each in Michigan and Nevada. To prepare for the field tests, two staff developers from each site attended a one-week summer institute, to familiarize themselves with the guides and engage in sample workshop activities. During the subsequent academic year, the staff developers carried out workshops at their home sites to introduce the guides to teachers in their regions. These workshops lasted from two to three hours per topic. From among the workshop participants, the staff developers recruited teachers to field-test the *Stuff That Works!* activities in their own classrooms and to evaluate the guides. Data from these field tests then became the basis for major revisions that are reflected in the current versions of all five guides.

Chapter 1

APPETIZERS

the worst thing about empty packages and containers is also the best thing: they are everywhere. Discarded packages can be fascinating and instructive, and are often full of mysteries. To start exploring packaging, just look around for some. Look in the garbage, the basement, the street, the park, and the cafeteria. Search for interesting jars, jugs, bottles, bags, and boxes. If computer shipments or other large deliveries have been made recently, find out where the boxes are. Look for cushioning materials as well as cartons.

Scavenger Hunt 1: Package Designs

Let's begin with this scavenger hunt challenge: How many different package designs can you find in each of these categories?

Cardboard Boxes and Cartons

- Two-piece cartons with telescoping lids—lids that are slightly larger than the boxes (Figure 1-1)
- Folding boxes, which are held together by mechanical tabs and slots, and can be unfolded completely (Figure 1-2)
- "Gabletop" containers, commonly used for milk and juice, that open and reseal (Figure 1-3)
- Boxes that incorporate dispensers, such as Kleenex boxes (Figure 1-4)
- Boxes of unusual geometry, such as cylindrical, heart-shaped, triangular, trapezoidal, hexagonal, and octagonal boxes (Figure 1-5)
- Boxes that both contain and display the product (Figure 1-6)
- Boxes that include a pour spout (Figure 1-7)
- Pizza boxes of various designs

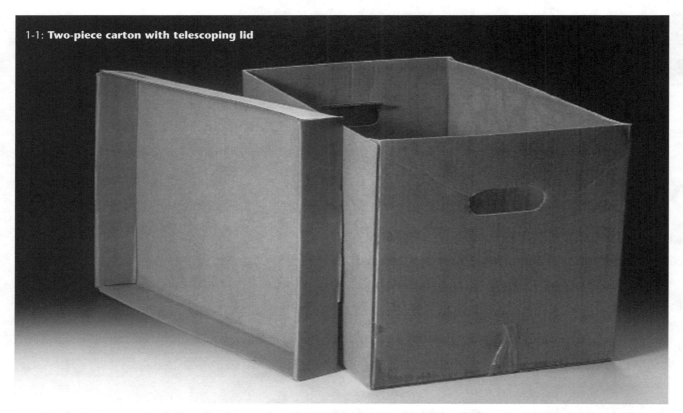

1-1: **Two-piece carton with telescoping lid**

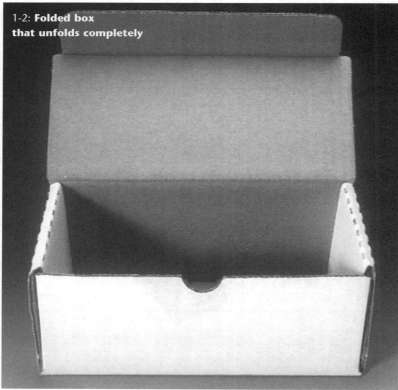

1-2: **Folded box that unfolds completely**

1-3: **"Gabletop" carton with built-in pour spout**

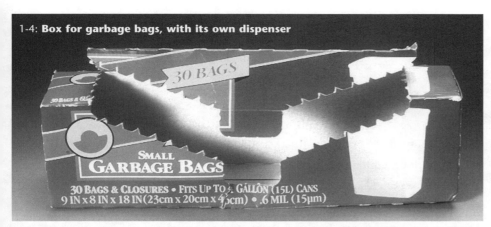

1-4: Box for garbage bags, with its own dispenser

1-5: Pyramid-shaped box

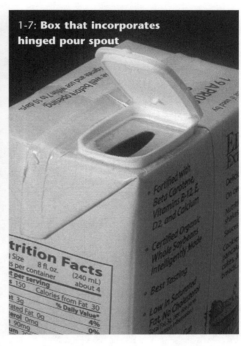

1-7: Box that incorporates hinged pour spout

1-6: Box for expensive perfume designed to display both bottle and sprayer

Plastic Containers

• Card-mounted "blister packs" that use both cardboard and plastic (Figure 1-8)
• Take-out food containers (Figure 1-9)
• Plastic boxes that keep the product wet or dry (Figure 1-10)

1-8: **Blister pack**

1-9: **Takeout container from an upscale restaurant**

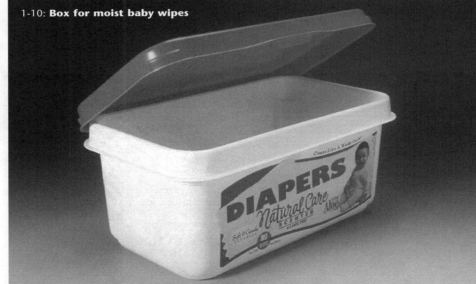

1-10: **Box for moist baby wipes**

Packaging Material Found Inside Boxes

- Materials used to make compartments, such as slotted cardboard partitions, plastic cookie trays, cardboard spacers, etc. (Figure 1-9)

- Materials used for cushioning, such as bubble wrap, peanuts, Styrofoam inserts, foam rubber, crepe paper, etc. (Figure 1-11)

1-11: **Assorted cushioning materials**

Bottles and Their Lids

- Different methods of child-proofing pharmaceutical items, including "push-then turn," "squeeze-then-turn," and "line-up-the-arrows" (Figure 1-12)
- Decorative glass and plastic bottles for perfumes and cosmetic items

- Plastic bottles for cleaning fluids, with molded built-in handles
- Shampoo containers of different types
- Squeeze bottles with different closure methods, such as dishwashing soap "push-pull" tops, hinged tops, and "flip spouts" (Figure 1-13)

- Pump dispensers (Figure 1-14)
- Spray dispensers (Figure 1-15)
- Dispensing methods other than squeeze, pump, and spray, such as roll-ons, "stick" dispensers, eyedroppers, brushes, daubers, etc.

1-12: **Child-proof tops**

1-13: **Assorted tops for squeeze bottles**

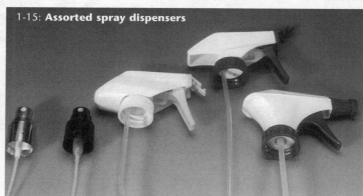

1-15: **Assorted spray dispensers**

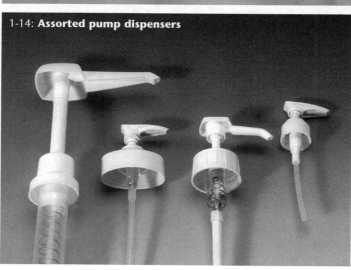

1-14: **Assorted pump dispensers**

Bags and Envelopes

- Shopping bags with the handles attached in different ways, such as cutouts, glued handles, glued patches holding handles, slots for string handles, etc. (Figure 1-16)
- Envelopes made in different ways
- Multi-layer bags, such as cellophane snack food bags with metallized layers
- Reclosable food bags with different closure methods, such as tab-in-pocket, zippers, and metal ties

1-16: **Shopping bags with an assortment of handles**

Examining Your Collection

As you look carefully at interesting examples of packaging, you will probably begin asking yourself questions like these:

- What type of cushioning is most effective for a particular product?
- Which pump or spray dispenser works best?
- What type of shopping bag or grocery bag is really the strongest?

These are all questions you can answer for yourself, as we shall see. There are a host of other questions that can help you organize and build your packaging collection:

- What set of problems was this package designed to solve?
- How was this package made?
- Why was one material used and not another?
- How well does it work for its purpose, compared with other designs?
- What else could this package or container be used for?
- How do the properties of the package match up with the properties of the product?

This last question is a subtle one. For example, spray dispensers and roll-ons are suited for thin, watery liquids; pump dispensers, for thick, viscous liquids; and stick dispensers, for waxy solids. What other kinds of correspondence can you find? At the end of the next chapter, we will help you with some of these issues by providing some of the technical background on packaging materials. In the meantime, we hope you will consider some of these issues for yourself. Above all, we urge you to look carefully at all of the packaging you come across.

Scavenger Hunt 2: Everyday Packaging Problems

Once you start, you will quickly accumulate a substantial collection of packages. You may also run out of space. Don't worry. This second scavenger hunt requires no space at all, except for a piece of paper. Use it to make a list of "everyday packaging problems"—the kind that crop up as nuisances in daily life. Here are some examples:

• How can I send these fragile objects through the mail?
• Which shopping bag should I use to take out the garbage?
• How am I going to get these heavy books to school? What if it starts raining on the way?
• How can I transport the food I just cooked without getting burned or without it leaking?

• This broken box is the only one I have left. How can I repair it?
• Is double bagging really necessary?

Some questions will be easy to answer just by looking at the packages in your collection. Others will require further investigation. Keep your list handy for future use. It can be a starting point for your work on packaging with children.

Packaging Mystery Challenges

Here are some "packaging puzzles" for you to solve. Some of them require that you work with a partner.

1. What was in this box?

Separately, you and your partner should each assemble a collection of interesting boxes of roughly equal size along with their contents. Remove all the contents and put the contents in one area and the packages in another. If the packages have identifying information, cover it over with dark paper or tape. The other person tries to match the contents with the package.

2. How was the cushioning material arranged?

Find a box with interesting cushioning material, such as the one in Figure 1-17. Remove the contents of the box and spread

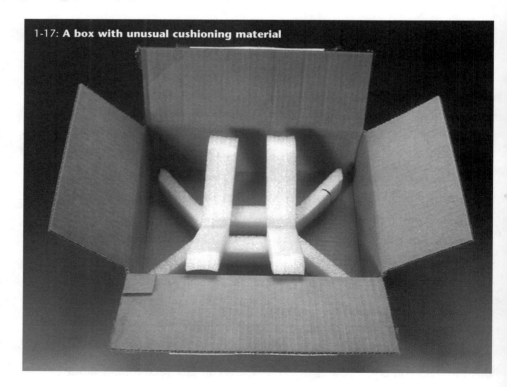

1-17: **A box with unusual cushioning material**

the cushioning material out on a table. Your partner has to figure out how the cushioning material was arranged in the box.

3. What would this closed box look like open?

From your collection, select a cardboard box that is folded into shape and held together without glue or tape, such as the one shown in Figure 1-18. Sketch what you think it would look like completely unfolded and laid out flat. (The answer for the box in Figure 1-18 is shown in Figure 1-19.) You can also play this game in reverse. Starting with an unfolded box, try to predict what the box would look like folded into shape.

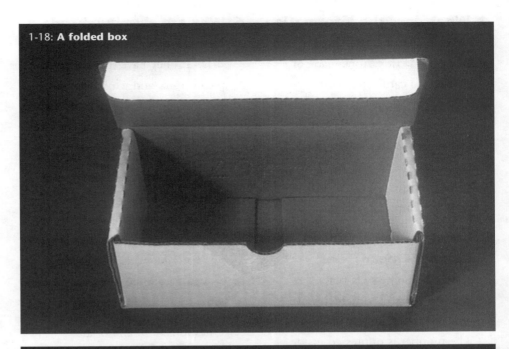

1-18: **A folded box**

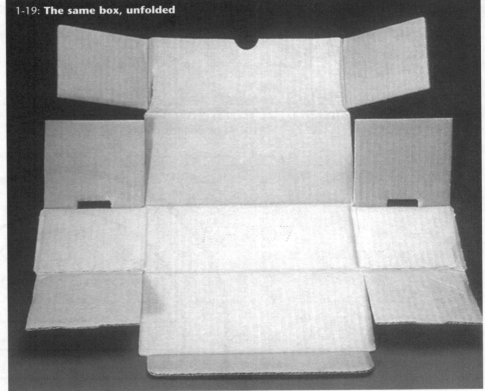

1-19: **The same box, unfolded**

4. What does the information in the "box certificate" mean? Most cartons made of corrugated cardboard have a little certificate printed somewhere on them. An example is shown in Figure 1-20. Do all of these certificates carry the same kind of information? What in the world does this information mean?

5. What do the recycling numbers mean?

Except for some soda bottles, nearly all plastic containers have a recycling number on them. The number is impressed inside a little triangle, usually on the bottom of the container. How many different numbers are used? What does each number mean? Why don't soda bottles usually have them?

6. Why are aluminum soda cans shaped like that? Why do they have that dome on the bottom, and why does the circumference get smaller at the top, compared with the rest of the can?

7. Can you think of a better way to package compact discs (CDs)?.

If you've bought one recently and had difficulty opening it or if you've ever had one of the hinges snap on you, it should be pretty easy to come up with a better design.

Questions 4, 5, and 6 are dealt with in detail in Appendix A. But first try to figure them out for yourself!

1-20: **A box certificate from a cardboard carton**

Chapter 2

CONCEPTS

What Are Structures?

Structures in Your Living Room, Kitchen, Bathroom, and Closet

In Chapter 1, we examined a wide variety of packaging materials, including boxes, bags, and bottles. Each of these is an example of a *structure*. Very simply, a structure is anything designed to hold something up or down, in or out, together or apart. Often people think of structures as very large things, such as bridges, buildings, and towers, but there are lots of structures that are much smaller, easier to understand, and more convenient to study. Discarded packaging materials are readily available and free! In this chapter, we deal first with the more general subject of structures, and then return to packaging as a particularly familiar and interesting category of the larger topic. We'll begin by looking for structures everywhere, just as we began by looking for packaging.

Here are some examples of structures you can find in your home:

- Cups, drinking glasses, plates, and bowls
- Luggage
- Furniture
- Picture frames
- Shoes
- Hooks
- Hangers
- Bars for holding shower curtains, towels, hangers, etc.
- Clotheslines
- Tripods and easels
- Light fixtures
- Umbrellas
- Pipes
- Ladders and step stools
- Pots and pans
- Cardboard tubes found inside toilet-paper and paper-towel rolls
- Anything held together with tape, glue, or string
- Anything that has been sewn, stitched, woven, braided, knitted, or knotted

Structures in Nature

Humans created all of the structures listed above, but we are not the only creatures that make structures. Some structures made by animals include:

- Hives
- Nests
- Spider webs
- Ant hills
- Beaver dams
- Burrows
- Dens and lairs
- Mole hills
- Coral reefs
- Cocoons

Plants and animals are themselves made of structures, and every organism has a long list of structural problems to contend with. A tree, an eggshell, and a fruit are examples of structures. Our own bodies are held up by a complex assortment of structural parts. Some of the structural elements in the human body include:

- Bones
- Muscles
- Tendons
- Ligaments
- Blood vessels
- Skin
- Teeth
- Nails
- Cell walls

Humpty Dumpty, and Other Tales of Structural Failure

How is it that there are so many structures within and around us, but that normally escape our notice? By definition, a structure is something that supports a load. To paraphrase a line from a movie, when a structure does its job, nobody even notices; but when it fails, there's a big mess! Many structures are not supposed to move at all, for example, light fixtures, hooks, and towel racks. Other structures are allowed to move, but only as a complete unit. If you move a chair from one side of the room to the other, it remains intact as a structure. Once the seat gives way, the back comes loose, or one of the legs breaks in half, the structure is no longer effective.

Folding structures, such as folding chairs and umbrellas, are special cases. Here, the parts of the structure are supposed to move, but only in a controlled way, and only when the structure is deliberately being folded and unfolded. At these times, the device is operating as a *mechanism*, not a structure. When someone is sitting on the chair or standing under the umbrella, the parts should not move. Under these circumstances, it is functioning as a structure.

The fact that the parts are not supposed to move is what makes structures difficult to analyze. A structure is composed of parts, just like a mechanism, but it is hard to see where one part ends and the next one begins, because they are not supposed to be "moving parts." When they do move, the structure is said to have failed. Structures become much easier to analyze and understand after they have failed because then you can see some of the parts and how they were once connected.

Structural failure must be of great interest to young children, because, as Petroski (1992) points out, many nursery rhymes deal with structure problems. These include Humpty Dumpty's mishap, Jack's accident descending the hill, Rock-a-Bye-Baby's fall, and the collapse of London Bridge. Adults are also intrigued by the failure of structures, although they would prefer them not to be close by. Much of the enjoyment of action movies, wild-ride documentaries, and slapstick comedies comes from watching cars, planes, buildings, and furniture undergo structural failure.

How Structures Work

Saved by the Stretch

When a structure fails, it is no longer able to resist the *forces* that are "trying" to make it give way. These forces are called *loads*. They include gravity, wind, earthquakes, people pushing or pulling, and impacts from other objects. A structure has to "fight back" against the loads that are working to make it fail. How does it do this?

Let's begin with a very simple structure. Take a rubber band, the longest you can find, and loop one end around your finger. Dangle a moderately heavy object, such as a stapler, from the other end (Figure 2-1). The rubber band will probably stretch a little bit, but more than likely, it will hold the stapler suspended. How does it do this?

The arrows at the right side of Figure 2-1 illustrate how. The downward arrow represents the force of gravity on the stapler. The rubber band is the structure; the force of gravity on the stapler is the load, represented by the downward arrow. This is the force responsible for stretching the rubber band. As the rubber band gets longer, it resists, pulling back. This "pulling back" is also a force; it is shown by the arrow pointing upward in Figure 2-1, labeled "Springiness." As the rubber band stretches more and more, the

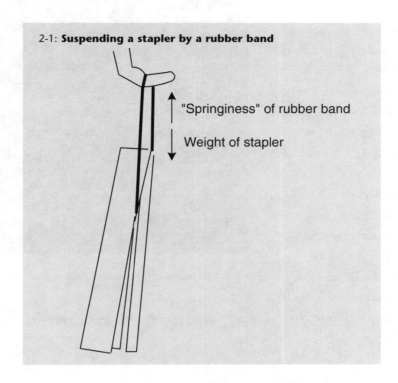

2-1: **Suspending a stapler by a rubber band**

"Springiness" of rubber band

Weight of stapler

"springiness" becomes greater and greater, until it exactly cancels the load due to gravity. At this point, the "tug-of-war" is a draw, the stapler comes to rest, and is said to have reached *equilibrium*.

Where does "springiness" come from? Anything we call "a solid" has fairly strong bonds between the atoms, which allow it to resist loads far better than a liquid or a gas. A rubber band resists being stretched because the atoms attract one another, and try to restore the material to its original shape. This "restoring force" is what "fights" the load, and makes a structure possible. You can actually see a rubber band stretch, but most materials do not lengthen enough for the change to be

visible. However, every material does actually get slightly longer when pulled, whether it is made of string, rope, or steel cable.

Loading something by pulling on it is called *tension*, which comes from the same root as "extend." Simple structures that work in tension are called *ties*. Some common examples of ties are fishing line, shower-curtain rings, luggage straps, yo-yo strings, ceiling fixtures, and elevator cables. A common experience with many ties is that they come loose at the points of attachment. This happens, for example, when a luggage strap tears away from itself near the little clip that is attached to the bag (Figure 2-2).

Stacks Are Structures

Someplace in your house or classroom, you can probably find a pile of things, such as books, papers, CD's, blocks, or boxes. A stack of things is a very simple kind of structure, in which the items near the top are held up by the ones lower down. Figure 2-3 shows the simplest possible stack, where block A is sitting on top of block B.

Notice from the diagram that block B is slightly shorter and wider than block A. Before they were stacked, block B looked exactly like block A, but putting a block on top of it made B very slightly shorter and wider. In other words, B is *compressed* by the weight of A above it. The change in shape of B is exaggerated in the drawing, unless B is made of something really flexible, like foam rubber or Jell-O. Most materials would be deformed much less than this, but every material gets compressed at least a tiny bit, even if it is really stiff, like steel, concrete, or stone.

In this example, block B is the structure, and block A is the load. As a result of being compressed like a spring, block B tries to return to its original position. In doing so, it pushes back up on block A. When the upward-pointing "springiness" exactly balances the downward-pointing load, the two forces cancel, and the system is in equilibrium. The block on the

2-2: **Shoulder bag strap comming loose at point of attachment**

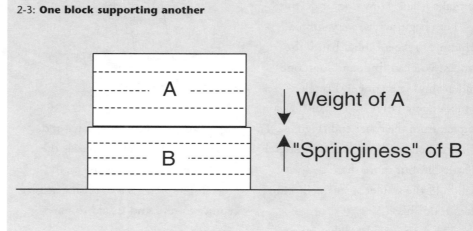

2-3: **One block supporting another**

A

B

↓ Weight of A

↑ "Springiness" of B

bottom is loaded in *compression*, because the load compresses it slightly. As in tension loading, the forces inside the structure work to resist the change from its normal shape. In the rubber band, the forces came from the attraction among atoms being pulled too far apart. In compression, the atoms are being pushed too close together, and their pushing back apart is what makes the structure work.

Stacking is not the most efficient or reliable way to make a *compression structure*. For one thing, the items have to have fairly flat tops and bottoms, or the whole structure may topple. Also, stacking may not provide much resistance against a push from the side. Worse yet, stacking does not use material efficiently. A chair, for example, could be made by stacking material all the way from the seat down to the floor, as shown on the left side of Figure 2-4. However, most chairs don't use this type of design. Instead, they use only four legs to support the load, as shown on the right side of Figure 2-4, and eliminate the rest of the material within the dashed lines. The four legs can provide enough resistance in compression, but with much less material.

Simple compression elements, such as chair legs, are called *struts*. Other examples of compression elements are the central pole of an umbrella, the legs of a tripod or easel, the vertical parts of a bicycle frame, the side pieces of a bookshelf or ladder, columns in a building, and the pole of a floor lamp or fan. Some structures can be supported using either struts or ties. Figure 2-5 shows an example. The left side shows a side view of a shelf, supported from below by a strut. On the right side, a similar shelf is suspended from a tie. The arrows show the forces of compression in the strut and tension in the tie. Which method—struts or ties—works better? Ties have the disadvantage of requiring special connections to other parts of the structure, while struts are at least partly self-supporting, thanks to gravity. Furthermore, ties don't usually create as rigid a structure as struts. In Figure 2-5, the suspended shelf on the right could fold upwards slightly, while the strut-supported shelf on the right has much less room for movement.

On the other hand, struts also have their disadvantages. The taller a strut is, the wider it has to be to avoid the

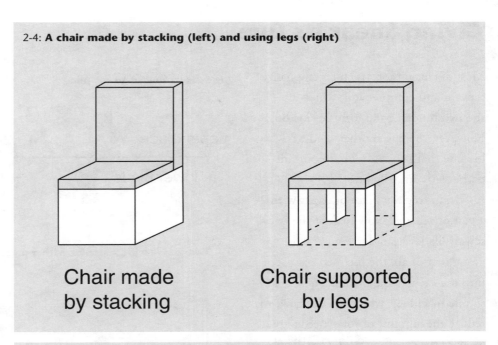

2-4: **A chair made by stacking (left) and using legs (right)**

Chair made
by stacking

Chair supported
by legs

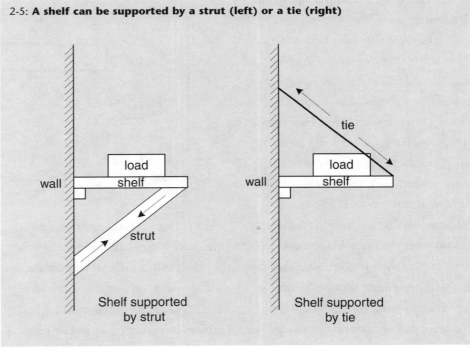

2-5: **A shelf can be supported by a strut (left) or a tie (right)**

wall — load / shelf — strut — Shelf supported by strut

wall — load / shelf — tie — Shelf supported by tie

chance of buckling, as we shall see. The trunks of tall trees have much larger diameters than short ones. Ties don't have this problem. A piece of string, fishing line, wire, chain, or spider web material will support the same amount of weight regardless of how long it is. Vogel (1998) points out that nature tends to prefer ties, while human designers are more likely to go with struts.

Giving Shear Its Due

Most discussions of structures emphasize tension and compression, which are the two fundamental forms of loading. However, there is another kind of loading that is often more important, particularly in daily life.

Begin to cut a piece of paper with a pair of scissors. Now look at the two scissor blades head on (Figure 2-6). Note, first of all, the inclined plane cut into the each of the blades. These are simple machines, which are designed to reduce the amount of force needed to make the cut. (See **Stuff That Works!**, *Mechanisms and Other Systems.*)

Next, notice how the blades move, as shown by the two arrows in the diagram. At first sight, this looks like compression, because the blades are moving towards each other. However, compression would not result in cutting the paper. Note that the arrows don't point towards each other, as they would have to in compression. The upper blade moves down somewhat to the left of where the lower blade moves up. This combination is called *shear*, which is easy to remember because "shears" is just another word for scissors. When you use a pair of scissors to cut a piece of paper, you are applying a shear load that is more than the paper can resist. If the paper could resist the load, it would stay intact. Tearing is another way to overcome the paper's shear resistance.

Many structures have to support loads in shear, and therefore need to be able to resist shear. How do

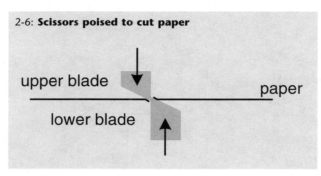

2-6: **Scissors poised to cut paper**

upper blade

lower blade

paper

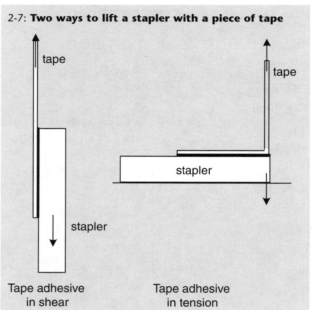

2-7: **Two ways to lift a stapler with a piece of tape**

tape

stapler

Tape adhesive in shear

tape

stapler

Tape adhesive in tension

shear resistance and tension resistance compare? Here is an experiment you can do, using a little tape and a stapler, or a similar object of about the same weight, to compare the shear and tension resistance of a piece of tape.

Cut a piece of masking tape or cellophane tape about six inches long. Press a few inches of the tape firmly on the flat side of the stapler, leaving a few inches of tape extending beyond it. Now hold the free end of the tape vertically, and lift the stapler with it, as shown on the left side of Figure 2-7. The tape will likely support the stapler.

If not, simply press the tape more firmly, or reposition it so more of its surface is in contact with the stapler. Next, place the stapler on a table, so it is lying on the side opposite the tape, as shown on the right side of Figure 2-7. Again, try to lift the stapler with the tape. More than likely, the tape will pull off, leaving the stapler behind.

Why? A piece of tape is simply a strip of paper with an adhesive on one side. In Figure 2-7, the heavy black line represents the adhesive. In the first part of the experiment, the pull on the tape, represented by the upward-pointing arrow, is to the left of the downward-pointing arrow that symbolizes the weight of the stapler. As with the scissors, the two forces are not aligned. They operate on the left and right sides of the adhesive coating of the tape, which is therefore loaded in shear. The tape holds because the adhesive has good shear resistance.

Now, let's look at what happened in the second part of the experiment. The force up on the adhesive was directly above the downward weight of part of the stapler, as shown on the right side of Figure 2-7. This time, the adhesive was loaded in tension, which it cannot resist very well, and it gave way. As a result, the tape lifted off the stapler because the adhesive couldn't resist the tension.

In discussing ties, we noted that one disadvantage of these structures is the problem of making secure attachments. Ties don't usually fail in tension, because strings, ropes, chains, straps, and wires are pretty strong. They usually separate at the points of attachment, and shear is nearly always the culprit. The luggage strap in Figure 2-2, for example, didn't fail in tension. The strap was originally looped around the clip, and then sewn to itself, as shown in Figure 2-8. When the weight

of the bag was too great, this pulled down on the short end of the strap; meanwhile the long end was held in place by someone's shoulder. This combination set up shear forces on either side of the stitched attachment, shown by the heavy arrows. These shear forces made the strap separate from itself at the stitching.

Shear is at its worst where a part protrudes from the main body, as in the hinges of a CD case; or where a protruding part is attached to a flat base, as with a car's side view mirror. Shear failure accounts for many annoying little problems:

- Broken hinges of CD cases (Figure 2-9)
- Broken handles of pots, pans, cups, and pitchers
- Buttons and snaps that pop off clothing because the thread unravels
- Wires and cables that pull out of computers, stereos, and other electronic gear
- Knobs that pull off their shafts
- Side-view mirrors that come off cars
- Door stops that separate from the wall
- Shoelaces that tear out the side of the shoelace hole (Figure 2-10), etc.

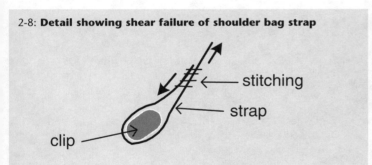

2-8: Detail showing shear failure of shoulder bag strap

stitching

strap

clip

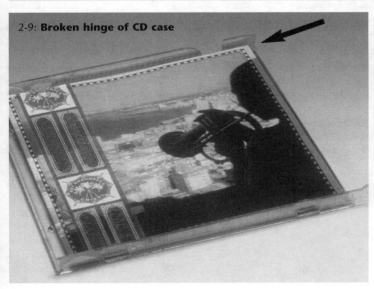

2-9: Broken hinge of CD case

2-10: Shoelace that has torn through the side of the hole

Common demonstrations of tension and compression involve two people supporting each other in two different ways. In the demonstration of compression, the two lean towards one another and support each other by pushing their palms together (Figure 2-11). If you do this experiment with another person, try to pick someone roughly the same size as you. As you lean together, you will feel the compressive forces in your arms.

The other demonstration, used as an example of tension, has the two people leaning away from each other and supporting one another by pulling on one another's hands (Figure 2-12). Again, you should do this with someone roughly the same size. This one is a little trickier, so lean back slowly until you are both sure of your footing. Sure enough, when you are in position, you can clearly feel the tension in your arms.

However, the demonstration in Figure 2-12 actually involves more than tension. As with all tension structures, there has to be a point of attachment, which is loaded in shear. In the demonstration shown in Figure 2-12, the attachment point is where the two hands meet. The hands are attached by having each person curl his or her fingers around the other person's, as shown in Figure 2-13. As they lean back, not only do they feel tension in their arms, but also a force in each hand that is trying to unravel the fingers. As the photograph shows, this force is a shear force. The so-called "tension demonstration" is really a demonstration of shear as well as tension.

2-11: **A Two-person structure, arms loaded in compression**

2-12: **A two-person structure, arms loaded in tension**

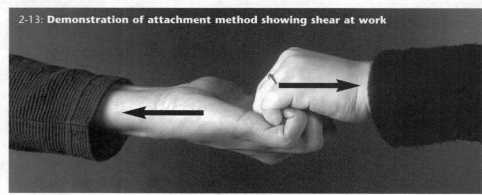

2-13: **Demonstration of attachment method showing shear at work**

Fascinating Fasteners

Some of the most important parts of structures are fasteners, which are devices specially designed to hold parts together. Fasteners are widely used for packaging, construction, decorating, and keeping things organized. The most commonly used fasteners for joining wood are nails and screws. Wood can also be joined using glue. For paper or cardboard, some fastener choices are thumbtacks, paper clips, and staples. Another possibility is to use some kind of spring clip, such as a "bulldog" clip, the clip found on a clipboard, or the "press" clips inside some binders. Other binders use loose-leaf rings or spiral-shaped wire. Glue or tape can be used as well. Figuring out the best way to keep papers together is a structures problem.

We will focus on how the loose-leaf ring, the paper clip, staple, and binder clip work to hold papers together. What kinds of loading do they resist well or not so well? To answer this question, suppose you need to join two sheets of paper together at one corner. A tension load would try to lift the top sheet vertically, while shear loading would tend to make one page slide horizontally over the other (Figure 2-14).

Now, suppose the two sheets are joined by a loose-leaf ring (Figure 2-15, left). The ring does not prevent the top piece from being lifted off the bottom piece. In other words, it offers no resist-

ance to tension. On the other hand, it does resist shear loading, because the metal binder prevents the pages from being shifted sideways very far. The page can move horizontally only by tearing, which can be a problem with loose-leafs. What happens in this case is similar to the shoelace tear-out problem shown in Figure 2-10. If the paper tears, it's because the force exceeds the shear strength of the paper itself.

The situation with the paper clip is exactly the opposite. It offers little resistance to sliding motion, because the top piece could easily slip out from under the paper clip; but it would not easily permit the top piece to lift right off (Figure 2-15, right). The loose-leaf ring passes vertically through the paper, so it resists horizontal loading. The paper clip, on the other hand, has horizontal wire loops that prevent vertical but not horizontal movement.

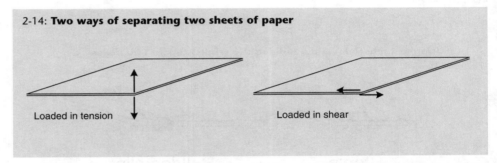

2-14: **Two ways of separating two sheets of paper**

Loaded in tension   Loaded in shear

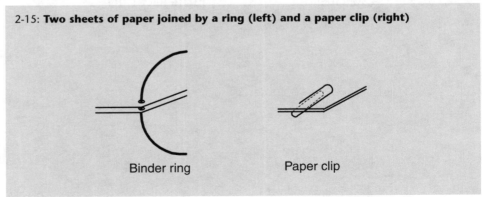

2-15: **Two sheets of paper joined by a ring (left) and a paper clip (right)**

Binder ring   Paper clip

Next, consider what a staple does. The two papers can move neither vertically nor horizontally because the staple offers resistance against both shear and tension. It is sort of like a combination of a binder ring and a paper clip, running first horizontally, then vertically, and then horizontally again. Staples are also very cheap. Their major drawback is that they require a special tool (a stapler) to insert them and form them into their remarkable shape. A spring clip also provides both kinds of resistance, but for a different reason. The spring exerts enough force to prevent the two sheets from moving either vertically or horizontally (Figure 2-16).

The most common fasteners for joining wood are the nail and the screw. Suppose a nail joins two boards, one on top of the other. Like a loose-leaf ring, its vertical shaft prevents sideways motion, and it therefore resists shear. However, a nail is fairly easy to pull out, so its resistance to tension is not so good. A screw, on the other hand, permits neither horizontal nor vertical movement, because the threads prevent it from being pulled out easily. Like a staple, it resists both tension and shear.

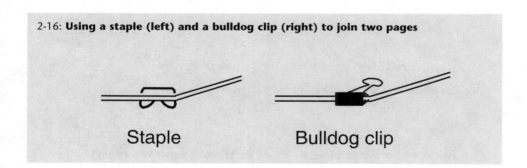

2-16: **Using a staple (left) and a bulldog clip (right) to join two pages**

Staple   Bulldog clip

Problems with Shelves, and with Beams in General

Are you always short of storage space? Shelves provide an easy solution to many common storage problems, and are a very useful kind of structure. If you don't have enough storage space, you can make your own shelves using tape and recycled cardboard (Figure 2-17). There are some obvious problems with this design, but they can easily be fixed. Homemade shelves may not look as nice as the manufactured variety, but they are easier to repair, and the materials are mostly free!

A structure like a shelf, which is held in place at either end, and loaded in the middle, is called a *beam*. Other examples of beams include most table-tops, chair seats, ladder rungs, and slats for holding up mattresses or bedsprings. The loading of a beam combines all of the types we have been discussing: tension, compression, and shear.

Let's look at what happened to the shelves in Figure 2-17. At A, two shelves on the left have broken free of their supports. Here is a typical attachment problem, in which the original

tape support has failed. Next, notice the shelf on the right (B) that is sagging under its load. This shelf is supported at both ends, but the weight has distorted its shape, especially near the middle. This shelf is experiencing both tension and compression.

We'll consider the attachment problem first. You can learn a lot about attachment issues just by looking at some commercially-made shelves to see how they are supported. Some common attachment methods are:

2-17: **Homemade shelves**

• screws or bolts, passing through the vertical side into the shelf, as are often found in steel shelving units;

• little brackets that hang the shelf from above, by means of slots cut into the sides, as exist in some storage cabinets;

• little pins or platforms that rest in holes on the sides, and support the shelves from below, as in most bookcases;

• one-piece molded construction, which forms the shelf out of the same piece as the side supports, as in most refrigerator door shelves.

Whatever method is used, the primary job of a shelf support is to resist shear, hopefully better than the failed supports in Figure 2-17.

Next, we'll turn to the bending problem. To clarify what happens to a shelf under load, Figure 2-18 shows a beam whose bending has been exaggerated.

There are several things to notice about the beam in the diagram. The supports experience maximum shear, because they have to resist the downward weight of the load. Notice also that as the beam bends, the part on top is forced to become shorter, while that on the bottom becomes longer. As a result, the upper half is loaded in compression, while the bottom half is loaded in tension. The maximum tension is along the very bottom surface, while the maximum compression is along the top surface, because those places are where the distortion is the greatest.

Here is a little experiment you can do to see how tension and compression make beams break. Grasp a small piece of wood, such as a craft stick, with one end in either hand. Now, push the middle with your thumbs until the wood breaks. Examine the broken halves carefully (Figure 2-19). The side your thumbs were on is like the top of the beam in Figure 2-18. Its wood fibers were in compression. The broken fibers on this side are short, and you can probably find some that folded back on themselves. On the side away from your thumbs, the wood fibers were loaded in tension. The long jagged edges are wood fibers that were originally connected, but pulled away from each other due to the tensile loading. Because these fibers are very strong in tension, they tend to fail by slipping away from each other, in shear, rather than by breaking, which happens on the compression side. That's why the jagged edges are so much longer on the tension side.

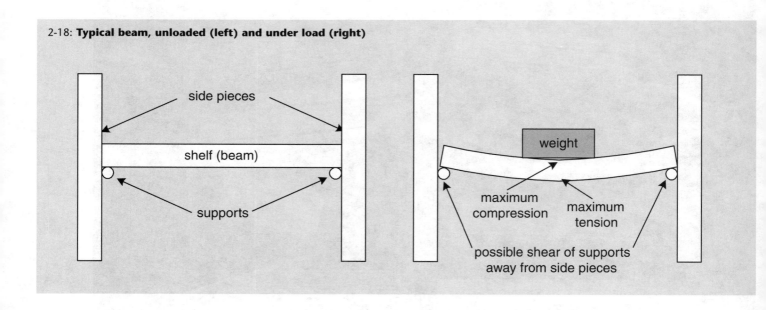

2-18: **Typical beam, unloaded (left) and under load (right)**

side pieces

shelf (beam)

supports

weight

maximum compression

maximum tension

possible shear of supports away from side pieces

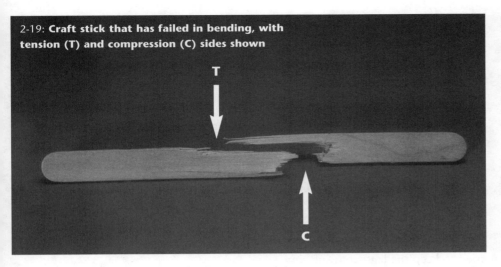

2-19: **Craft stick that has failed in bending, with tension (T) and compression (C) sides shown**

T

C

What Can Go Wrong in Compression

So far, we have looked at structural elements that have to resist tension or shear. In some ways, compressive loads are easier to work with. The earliest monuments built by humans, and the largest ones still standing, were built mostly by piling large blocks on one another. These include the Pyramids, the Great Wall of China, and the Roman aqueducts. Piling things on top of one another sounds simple enough, and it avoids the attachment problem, but large blocks are heavy and hard to move. Nowadays, most compression structures use much less material, as in the chair of Figure 2-4.

A vertical strut, much thinner than it is tall, is called a *column*. The legs of a table, chair, stool, or tripod and the sides of a bookshelf or a ladder are all columns. A big problem with all of these structures is that the columns have to remain more or less vertical. When the legs give way by slipping outward, they are said to *splay*.

Because the worst areas for compression and tension are at the top and bottom, respectively, these are the parts of a beam that need to have the greatest resistance. The center of the beam doesn't need to do nearly as much. In the construction industry, the most common beams have a cross-section shaped like the letter "I" and are called *I-beams*. Figure 2-20 (left) shows the end view of an I-beam. Most of the material is contained in the top and bottom flanges. The slender center, called the *web*, serves mostly to join the flanges, which offer nearly all of the resistance to compression and tension.

A similar strategy is used in making cardboard cartons. Most cardboard used in packaging is corrugated, and has the cross-section shown in Figure 2-20 (right). The flat top and bottom, which are called *facings*, are made of cardboard that is nearly twice as heavy as the corrugated part in the middle, which is called the *medium*. The facings are like the top and bottom flanges of an I-beam; they offer most of the resistance to bending. The medium is similar to the web; its job is mostly to hold the top and bottom together, using as little material as necessary.

2-20: **An I-beam (left) and a piece of corrugated cardboard (right)**

flanges   web   medium →   ←— facings

"I" beam   Corrugated cardboard

Ladders have horizontal ties to keep the two sides from splaying (Figure 2-21). Most chairs, tables, tripods, and easels use techniques of one kind or another to prevent splaying. These include:

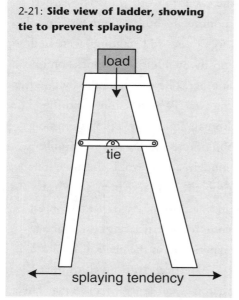

2-21: **Side view of ladder, showing tie to prevent splaying**

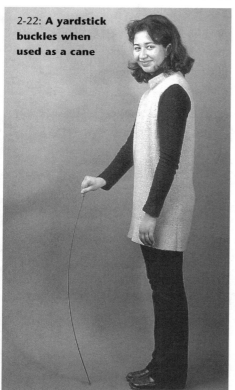

2-22: **A yardstick buckles when used as a cane**

• "X" shaped cross braces to prevent sideways movement;
• slots that hold the legs captive under a tabletop or chair;
• chains or straps that work as ties to keep the legs of a tripod or easel in place;
• rings around the outside of stool legs, holding them together.

It's worth looking at a few chairs and tables to see how the designers tried to keep the legs from splaying.

A second problem with columns is that they can bend, just like beams. When a column bends under load, the problem is called *buckling*. Figure 2-22 shows why a yardstick does not make a good cane: it buckles under the weight of the user. To prevent buckling, a column needs to have a lot of material on its outsides. In fact, the best shape for buckling resistance is about the same as for a beam. For that reason, I-beams, stood on end, are used to make columns in buildings, and corrugated cardboard is used for the sides of cartons, as well as for the tops and bottoms.

However, there is one very big difference between bending and buckling. A beam can bend a little, and still support a load, as the sagging shelf does in Figure 2-17. Bending a little can actually make the beam stronger. On the other hand, when a column buckles, it's usually all over. Even a very slight amount of buckling makes a column weaker, so it buckles more and more and gives way almost instantly!

The ancient Romans came up with a clever answer to the problems of building with beams and columns. They fashioned a beam and two columns out of one piece of material, and rounded it on the inside. The resulting structure is called an *arch* (Figure 2-23). It is a structure that works almost entirely in compression. There is an arch inside each of your feet. Together, your two arches have to support your entire body weight. Related to the arch, and even stronger, is the structure you get by rotating an arch around its center: the *dome*. Both the top and bottom of an eggshell are domes, which make an egg very difficult to break at the ends.

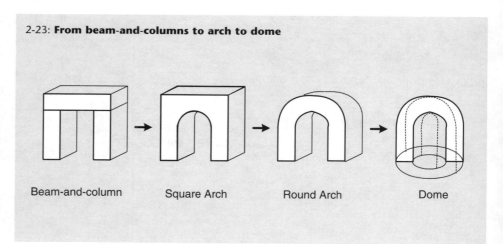

2-23: **From beam-and-columns to arch to dome**

Beam-and-column   Square Arch   Round Arch   Dome

Exploring Packaging

Materials You Can Find Anyplace

Packaging is everywhere. It encloses nearly everything people buy, carry with them, send to one another, and store. Sometimes, these packages work well, in which case they are barely noticed. Often, packaging fails in one way or another, at which point it becomes an annoyance. In either case, people rarely stop to examine or think about this extraordinarily important and common branch of technology.

Packaging and containers are not usually included in what people think of as technology. In a thought-provoking argument, Lewis Mumford (1967) suggests that this omission reflects a male bias. Containers are associated with traditionally female occupations, such as cooking, brewing, and gardening. Tools and weapons, in contrast, are more related to typically male pursuits like hunting, metalworking, and tool-making.

Although packaging issues are largely beneath conscious notice, at times they assume great importance, for both children and adults. Many children are very particular about the styles of bookbags or water bottles

they will take with them, while adults have strong preferences and beliefs in selecting shopping bags, boxes, and luggage. These beliefs are generally untested. Do you really know which of your shopping bags is strongest?

Most people take containers for granted, but they are taken very seriously by major manufacturers. The sales volume of corrugated cardboard boxes is considered a barometer for the health of the entire economy. Approximately 100 billion jars and bottles, and their caps or tops, are sold in the U.S. annually. Another 100 billion beverage cans change hands. In the soft-drink industry, the package is considered more important than the beverage it contains. Both the aluminum "pop-top" can and the plastic bottle were considered major, highly profitable innovations in the soda business. In many parts of the world, soda is transferred to a plastic bag before being sold to the customer. The can or bottle is simply too valuable to give away!

Besides being a major industry, packaging is a fertile curriculum

area. The materials are free, consisting of discarded items. They are very familiar to children and adults. Packaging problems abound in our daily lives, which provide a multitude of opportunities to analyze and redesign existing packages, and design new ones. These activities offer a host of connections with other areas of the curriculum, including language arts, social studies, science, and math.

Children enjoy writing about their discoveries and inventions related to packaging. The evolution of packaging and beliefs about it, offer a window into profound changes in our society. Analysis and design of boxes offers an engaging route into measurement, as well as plane and solid geometry. Product testing of packages invokes an array of science process issues, such as control of variables. The mechanical properties of packaging raise basic issues of structures, such as stability, compression, tension, shear and energy absorption. Packaging can be a vehicle for curriculum integration.

What Good Are Packages?

Nearly every commercial package is a carefully engineered product, cleverly designed to solve an impressive set of problems. A useful starting point in analyzing a package is: "What was this package designed to do?" Each package has a different set of tasks to fulfill, which is why there are so many different kinds of packages. Here are the kinds of needs that packages are supposed to fulfill:

1. **The package has to contain the product.** The most obvious function of a package is to keep the contents in one place and (if it is solid) in one piece. The most apparent packaging failures occur when the package seems about to stop keeping the product inside (Figure 2-24). Containment of the product can be complicated by conditions inside and outside the package. The plastic soda bottle, for example, has to withstand inside pressures reaching 5 or 6 times atmospheric, at temperatures ranging from below freezing to well above 100°F.

2. **The package should maintain its shape during shipping and storage.** Shoppers are usually reluctant to buy dented cans, although the product may not have been affected in any way. In addition, cans and cartons are often stacked high on one another for

2-24: **Carton that barely holds its contents**

storage, shipment, or display. The ones near the bottom have to support the weight of those above them. They need to be strong enough in compression to do so, or the entire stack may fall. Of course, some packages are not expected to keep their shape— for example, bags.

3. **The package may need to protect the contents from the environment.** This requirement is particularly important for food and beverages, which can spoil in a variety of ways. A small juice carton, which is shipped and stored at room temperature, has layers of plastic, paper, ink, and aluminum, each with a different purpose. To see this for yourself, cut an empty juice box open and pull apart the layers. The paper makes the box rigid, and holds the ink; the ink provides decoration and information; the aluminum keeps out light, oxygen, and microorganisms; and the plastic keeps the liquid from oozing out. Until this carton was developed, juice could not be sold in unrefrigerated boxes, because it had an unfortunate tendency to ferment and explode!

Often the need to protect the contents conflicts with the need for access (see #4, below). The small juice container is intended for one-time use, so reclosure is not an issue. However, many packages are too big for their contents to be consumed all at once. Many cracker and dry cereal packages use the "bag-in-box" strategy, which features a reclosable inner bag to keep out moisture, because atmospheric humidity can lead to a loss of crispness. Sometimes the cardboard top is also equipped with a reclosable tab-in-slot.

4. **The package should permit access to the product.** A container is useless if it is too difficult to open. The earliest metal cans, which were made of heavy iron, were normally opened with a hammer and chisel. Some modern packages seem pretty hard to open, but others are equipped with convenient dispensing devices, such as squeeze tops, straws, pump and spray dispensers, measuring cups, one-at-a-time tissue paper dispensers, etc.

5. **Some packages have to control access to product.** Access is not necessarily for everybody. Many foods and drugs now come in "tamper-evident" packages, which signal the

shopper if they have already been opened. There are a variety of ways of doing this: outer seals, inner seals, tear bands, mechanical "breakaway" seals (Figure 2-25), and vacuum buttons are all common. There are also "child-resistant" tops, used most commonly in packaging medicines. Common technologies include the "push-then-turn" top, the "squeeze-then-turn" top, and the "line-up-the arrows" top (Figure 2-26). These have to be hard enough to open that small children won't get into them, but not so hard that elderly or infirm patients are kept out.

6. **"Point-of-sale" packages (those displayed in stores) usually promote the product.** Before modern packaging was developed, goods were delivered to stores in bulk barrels and drums, and the storekeeper had to help with each purchase. Packaging made it possible for shoppers to select items themselves, in the self-service market,

now called the supermarket. On a typical hour-long visit to the supermarket, you see about 30,000 different products, or about 10 per second. All of them are competing for your attention. Marketing professionals regard the package as the "punch line" of a marketing campaign—their very last chance to grab your attention before you make that final choice. A tremendous amount of time and expense goes into the graphic design of those packages. This aspect of packaging is covered in the *Stuff That Works!* curriculum guide, *Signs, Symbols, and Codes.*

7. **Most packages also provide information.** There are many kinds of information that might appear on a package. Some of this information, such as the box certificate (see Figure 1-20 in Chapter 1) and the bar code, is not directed towards the consumer. The box certificate contains a message from the box manufacturer to the shipper. The bar code

2-25: **Tamper-evident "breakaway" seal on a bottle**

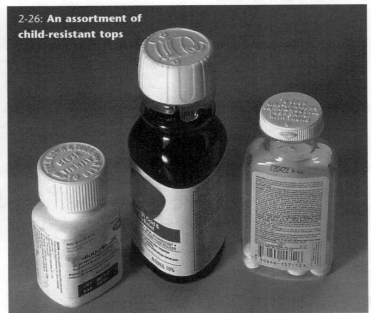

2-26: **An assortment of child-resistant tops**

is read by a computer and winds up in the store's database. The "Nutrition Facts" box, required by the U.S. Government on most food items, does provide information for the consumer. Many packages also have instructions, recipes, phone numbers, and other information for the user who cares to look. Recycling numbers, discussed in Appendix A, provide information to the consumer as well as to others involved in the recycling process.

8. **The package should have minimal environmental consequences to humans and other organisms.** The harm caused by certain packaging materials has led to the elimination of some technologies. Some early attempts to make plastic

soda bottles failed because they were suspected of releasing toxic by-products. The earliest pop-tops had a removable tab that was both a source of litter and also a health hazard to birds, fish, and barefoot humans. As a result, the removable top was replaced by one which remains with the can (Figure 2-27).

Looking at this list, it is obvious that most packages have more than one set of requirements to satisfy. The answer to "What was this package designed to do?" hardly ever has only one answer. On the other hand, some criteria are more important than others. For example, if a package can't hold its contents, it hardly matters what kind of information it conveys!

2-27: **Non-removable pop-top**

2-28: **Blister pack**

$\mathcal{A}$ Feeling for Packages: Surveying Strengths and Weaknesses

Once the criteria for evaluating a package have been established, it is reasonable to ask: "How well does the package meet these criteria?" In looking at any package, there will be some obvious ways in which it works or doesn't work.

Every package has been designed by someone to solve a problem. It may also create new problems. Those horribly hard-to-open "blister packs" (Figure 2-28) are good examples. Most blister packages hold the product with an over-

sized card in a clear bubble of plastic. The purposes of these packages are to:

- Make the product and promotional material visible;
- Provide a large surface area, making the package harder to steal;
- Make it difficult to remove the product from the package, further discouraging shoplifting.

These last two design goals are examples of controlling access. They

are also what make these packages so irritating—the product is hard to remove from the package, even when it has been legally acquired! Some styles of blister packages try to overcome this problem by providing perforated doors in the back of the cardboard, or by allowing the cardboard to slide off of the plastic top, or by providing a tab or break for easy opening.

Another example of problematic design is the "gabletop" container, which is used for milk, juice, and other liquid products. These containers are made of paper coated with plastic on both the inside and outside, sometimes with additional layers of plastic and aluminum foil inside. A quart or half-gallon is rarely consumed at one sitting, so these packages need to be resealable. At the same time, milk is easily spoiled, and the container has to provide a good barrier to light and oxygen, even after it has been opened the first time.

As with many other packages, the gabletops present a conflict between the goals of protection and access. The conventional milk container is opened by pulling both sides of the gable back until they are flat, and then opening them in the reverse direction so a pour spout forms. This can be very hard to do, especially for small children. Opening these cartons often leaves behind torn corners and ragged edges, which make it hard to pour the contents without dribbling (Figure 2-29). Probably as a result of these problems, some dairies and juice companies have come out with gabletop containers with screw-off tops (Figure 1-3 in Chapter 1). These usually have an inner seal, which is removed the first time the container is used.

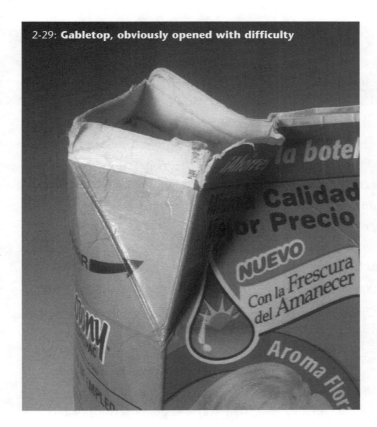

2-29: **Gabletop, obviously opened with difficulty**

How Does It Measure Up?

The question "How well does it work?" is pretty vague. There is rarely only one way to solve a packaging problem. There are nearly always alternative packages that are designed to do more or less the same thing. The question "How well does it work?" can be reformulated as "How does this package compare with alternatives?"

Comparing designs is a familiar task from science education, where units on product testing are intended to teach control of variables and fair testing. Product testing of packages arises from everyday questions: Which bag is strongest? Which cushioning material works best? Which pump dispenser is fastest? These issues come up frequently and the answers are often of practical importance.

Bags

Bag testing is an easy-to-do activity, which yields nearly immediate results. Add weights to each bag gradually, until it breaks. Some shopping bags are pretty strong, so you may need a lot of weight to break them. Weightlifting weights are useful for this purpose—they are heavy, have no sharp corners, and have the amount of weight stamped right on them.

What are the variables that need to be controlled in this experiment? One is the way the bags are supported; another is the way the weights are added. If they are dropped rather than carefully placed in the bag, the impact will have a greater effect than the gentle addition of weight. Weights with sharp edges or corners can puncture the side of the bag, creating a new area of weakness.

Weightlifting weights are not always the best type of weight for bag testing. They can pose a safety hazard with younger children, and are also difficult to transport. As an alternative, some teachers have used plastic one- or two-liter soda bottles filled with water. If a bottle drops on someone's foot, it will not cause injury. A liter of water weighs one kilogram, so this type of weight can also lead to a discussion of the metric system. Also, empty soda bottles are readily available and free, and can be transported empty, which makes it easier to bring them into the classroom. On the other hand, bottles

of water may not be heavy enough to break a bag. If you use these as weights, select relatively flimsy bags to test. Another strategy is to hold the bags by only one handle, which should require only half the weight, if the handles are the weakest links. With paper bags, you can also reduce the necessary weight by testing the bags wet rather than dry.

Testing results can be surprising. Bags often fail at the handles, so handle attachment methods can be very important (Figure 2-30). Most people assume that plastic bags are stronger than paper ones, but we have found paper shopping bags that hold well over 100 pounds. In Appendix A we discuss some of the different ways in which bags break.

Cushioning

Another fascinating area for product testing is comparing cushioning materials. These come in a wide variety of types. Inside of packages you can find spongy paper or cardboard padding; bubble wrap; Styrofoam sheets, blocks, end caps, "peanuts" or "figure eights" (Figure 2-31); crumpled or shredded newspaper; or even plastic bags filled with air (Figure 2-32).

These air-filled bags may be difficult to find, but you can make them yourself by blowing up plastic sandwich bags and tying them tight, like balloons. The messy wadding material used in book mailers consists of untreated paper fibers and bits of shredded newspaper.

2-30: **Shopping bag collection, showing many different handle types**

2-31: **Styrofoam "figure-eights"**

2-32: **Minimalism in cushioning: the air-filled bag**

To test cushioning material, you will need a standard fragile object, which the cushioning is supposed to protect. Most people think immediately of eggs, but these are problematic for several reasons. One is that they are very messy when they break. Also, raw eggs are sources of harmful bacteria, which can lead to food poisoning if hands aren't washed with hot soapy water. Also, eggs are no longer useable after they break, which means that food is wasted.

For these reasons, we have come up with several alternative to eggs as fragile products for testing cushioning. Some teachers have used water balloons, but these are messy too, and it is hard to get exactly the same volume each time. Another idea we had was to roll out thin lengths of modeling clay, and allowing it to dry without firing it. However, real modeling clay (not plasticene) is no longer so easy to find.

The best alternative we have found are snack items, such as very thin breadsticks, crackers, and chips, which are often found broken when the box is first opened. These can be wrapped in foil, wax paper, or paper towels to keep them edible after the test. Some other possibilities are chalk and mechanical pencil leads.

How should the test proceed? For control of variables, identical boxes should be used and identical amounts of cushioning material. To find out which cushioning technology is most effective, the boxes can be dropped from increasing heights, or increasing amounts of weight can be dropped onto them. These methods are not equivalent. Again, the results can be surprising. Some background on cushioning materials is presented in Appendix A.

Pump Dispensers

A third example of product testing involves pump dispensers. These come in a wide variety of types and sizes. You can find them on top of containers for hand soap, condiments, cleaning supplies. and other products (Figure 2-33). These fluids vary greatly in *viscosity*, or resistance to flow. Depending on the viscosity and the type of nozzle, the fluid may come out as a glob, a stream, or droplets. The amount that comes out each time will vary, depending on the dispenser as well as the fluid. Often, a pump dispenser needs to be primed a few times before any product will come out. There are a lot of issues here, which can be studied systematically through controlled product testing.

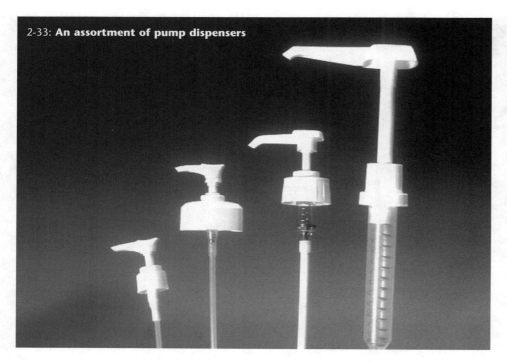

2-33: An assortment of pump dispensers

Package Design, Redesign, and Re-use

Having analyzed and tested a package, and discovered its weaknesses, it makes sense to ask: "How could this package be redesigned to do a better job?" If the package has already failed, the better question might be "How can this package be repaired, so it can still be used?" These two questions are not so different. Redesigning a package so it won't break usually involves reinforcing the very areas that are most likely to fail, before they actually do. "Repair" means doing pretty much the same thing after the failure has occurred.

To experiment with pump dispensers, it is not necessary to use the same fluid or the same container they came with. We have found it convenient to place the tube in an open pan of fluid. After experimenting with them for a while, children will think of their own criteria for evaluating their performance. In Christine Smith's sixth-grade class, the students came up with these performance criteria (see Chapter 4):

• How many times do you have to prime the pump before anything comes out?

• Once the fluid starts coming out, what is the average amount per stroke?

• How far does the stream of fluid travel?

One path of exploration is to test a variety of pump dispensers with the same fluid. Water is the most obvious choice. Alternatively, you can use the same pump dispenser and test a variety of fluids of different viscosities. Possible fluids include water, alcohol, mineral oil, corn syrup, etc. We examine how pump dispensers work in Appendix A.

One outcome of bag testing is that you will find yourself with lots of broken bags. They can break in a variety of ways, as we shall discuss in Appendix A. A pre-K/K class established a "Bag Repair Area" for fixing the broken bags. In the process, they learned how to identify how each bag had failed. Then they looked for ways to reinforce the bags so they wouldn't break so easily. These children were learning redesign at a very young age!

Another aspect of package design is to look for new uses for a package that has been discarded. Most environmental education programs focus on recycling as the solution to the solid waste problem, but re-use is both more beneficial to the environment and has more educational potential. Recycling is not so easy to learn about at an elementary level. For example, the recycling of plastic soda bottles into fleece pullover sweaters is a high-tech process involving a large investment and extensive technical know-how. Children can certainly collect soda bottles, but there is little else about recycling that is really accessible to them.

On the other hand, re-use of packaging material requires little or no expenditure, nor much prior background knowledge. For example, some sixth-grade classes decided to construct a model city out of re-used packaging materials. Their first step was to look for discarded materials, and think about or test the properties of what they found. Re-using old packages in new ways is a real challenge to the imagination. For example, the squeeze-dispenser top of a bottle of dishwashing detergent became the nose of a locomotive in the model city. Another example comes from a teacher workshop. One participant quietly walked over to the door and shut off the lights. On

the worktable, for all to see, was the lamp she had made from a discarded wire spool. These are two creative examples of re-use.

A third category of design involves creating something new. Of course no design is really "from scratch." A designer always brings knowledge of existing designs with her in thinking about a "new" design problem. For this reason, there is no clear-cut distinction between "redesign" and "new design." An example of a design problem that draws heavily on an existing design is the following:

> You will be provided with a folding box that can be unfolded flat and refolded, without tape or glue (Figure 2-34). Using the same basic design, draw, cut and construct a package for a block of a different size.

This problem requires considerable measurement, 2-D and 3-D geometry, and spatial visualization.

The most authentic type of design problem is one that arises naturally in

the course of children's lives. Sandra Skea's sixth-grade math students designed and made portable storage units to hold shoebox dioramas that had to be carried from class to class (see Chapter 4). A fifth-grade class designed see-through packages to both display and promote T-shirts they were selling as part of a school-wide poetry festival. A group of sixth-grade science students designed and made packages for fragile gift items they had made in class. Each of these projects began with design criteria that had arisen from a real need. Each group's design could be tested against these criteria, compared with the products of other groups, and redesigned and retested, if necessary.

All design projects can and should draw on existing designs. Looking carefully at someone else's design and drawing conclusions from it is what we call analysis. The connection between analysis and design happens when data from existing designs is used to inform a "new" design. Some teachers in a workshop decided to construct shelving units from discarded cardboard, tape, and string. They were unsure about how to support the shelves on the vertical side panels. To answer this question, they examined all of the cabinets, blackboards, desks, and bookshelves in the workshop area to find out how this problem had been solved in each of these units. They found about a dozen fundamentally different solutions, which provided many ideas for their own design problem.

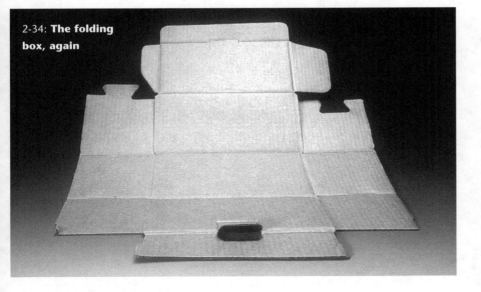

2-34: **The folding box, again**

Chapter 3
ACTIVITIES

the activities in this chapter are designed to give students direct experience with designing and evaluating different kinds of packaging and structures. The activities were created and tested by classroom teachers. Many of their experiences with these or similar activities are described in Chapter 4, "Stories."

Activities 1-7 deal with kinds of packaging materials and dispensing devices. Activities 8-13 deal with building and testing the strength of cardboard structures. The activities are designed to give students experience with many of the concepts discussed in Chapter 2, "Concepts."

All of the activities are correlated to standards in Science, Math, and English Language Arts. The standards are listed by number with each activity; the standards themselves are listed at the end of the chapter.

ACTIVITIES AT A GLANCE

Level	Activity Title	Page	What Students Learn About Packaging & Other Structures		
			Categorizing	Analyzing/Testing	Design/Redesign
Introductory	Exploring and Categorizing Packages	44	✗		
	Classifying Bags	48	✗		
	Packing a Bag	51		✗	✗
	It Fits Just Right	53			✗
Intermediate	How Strong Is This Bag?	54		✗	✗
	How Do You Package a Fragile Object?	57		✗	
	Which Pump Dispenser Works Best?	60		✗	
Advanced	How Does the Shape of a Column Affect Its Strength?	66		✗	
	How Does the Shape of a Shelf Affect Its Strength?	73		✗	✗
	How Does the Type of Cardboard Affect the Strength of a Shelf?	76		✗	✗
	How Does the Direction of the Corrugations Affect Its Strength of a Shelf?	78		✗	✗
	How Does the Type of Glue Affect the Strength of a Laminated Shelf?	81		✗	✗
	How Does the Support Method Affect the Strength of a Shelving Unit?	83		✗	✗

Activity № 1

Exploring and Categorizing Packages

Grade Level
K-6

Prerequisites
Understanding the concepts of categories and sorting

Overview
This is an early childhood activity that involves children in exploring and categorizing some of the many different kinds of packages they see in their daily lives.

Concepts
- Packaging comes in a wide variety of forms, sizes, shapes, materials, and functions
- Examples of packaging can be sorted into groups according to a variety of criteria

Vocabulary
- Handle
- Packaging
- Plastic
- Paper

Skills
- Observing, comparing, and contrasting differences in materials
- Classifying objects according to shape, function, composition, and other criteria
- Counting and measuring using manipulatives
- Listening attentively and speaking using appropriate vocabulary

Standards
- Benchmarks for Science Literacy: 1B
- Standards for the English Language Arts: 12
- National Science Education Standards: A
- Principles and Standards for School Mathematics: A1

Time Needed
Three or four periods

Materials
- Large assortment of clean, empty packaging materials such as boxes, cartons, plastic containers, paper and plastic bags, plastic bottles, and jugs
- Worksheet #1A (K-2) or #1B (3-6)

Pre-Activity Preparation
- Ask friends, family, and colleagues to help you collect examples of packaging. Place the items in an area of the classroom where students can see them. When students ask you about the collection, encourage them to speculate about what the items are and what they have in common.
- Always review safety issues regarding plastic bags and any other objects you think could pose a risk if mishandled.

Procedure, Grades K-2
1. Have a brainstorming session with the whole class about packaging. Record students' responses on an experience chart. Begin by asking such questions as:
   - "What do you think the word packaging means?"
   - "Can you name some examples of packaging?"
2. To review and reinforce students' understanding of sorting and creating categories, play "Guess My Rule." Gather a collection of various items, such as toys, counters, blocks, pencils, crayons, and so on. Sort them into groups (e.g., according to

color, use, material) and ask students to guess the "rule" for creating the groups. In other words, "What do the items in each group have in common?" To reinforce the concept, keep the categories simple and let the students add other items to the groups that match the criteria. Discuss the different ways objects can be sorted into groups.

3. Divide the class into groups of four and give each group several packaging items. Each group should have items that can be grouped together based on a variety of criteria—e.g., type of material (paper, plastic), kind of container (box, bag, bottle, jar), product contained in the package (food, cleaning supplies), and so on. Give the groups time to examine the items carefully and talk about what they see.

4. Ask the groups to organize their packaging materials into piles, according to the kinds of materials they are. Ask the groups to explain what's in each pile and why they're grouped together.

5. Distribute copies of Worksheet #1A to students. Ask them to draw one item from each grouping to identify the group. They then tally the number of examples of packaging items they put into that category. Work with younger students as necessary to help them record their findings.

Adaptation for Grades 3-6

1. As homework, ask students to bring examples of packaging materials from home.

2. With the whole class, discussing the meaning of "packaging" and list examples.

3. Divide students into cooperative learning groups and have them sort a collection of packaging items into categories. Encourage them to identify items that could belong to more than one category—e.g., a plastic juice container can be classified as "plastic," "bottle," and "beverage container."

4. Distribute Worksheet #1B. When students have completed the worksheet, bring the class together to share and discuss students' work.

5. As a follow-up homework assignment, ask students to find 10 examples of packaging at home or elsewhere, draw each one, and describe each package and what it is used for.

Worksheet #1A (Grades K-2)

Exploring and Classifying Packages

Name/Group   Date

Show your groups and how many are in each group.

Worksheet #1B (Grades 3-6)

Exploring and Classifying Packages

Name/Group _____ Date _____

What is packaging? _____

_____

_____

_____

_____

_____

List some different types of packaging things you know. _____

_____

_____

_____

_____

_____

_____

List the categories into which the packaging items in your collection can be sorted. Under each category,

describe the items that belong in that category. _____

_____

_____

_____

_____

_____

_____

Activity № 2

Classifying Bags

Grade Level
2-4

Prerequisites
- Basic knowledge of categorizing principles
- Knowledge of using charts and graphs to display data

Overview
This activity builds on Activity #1: "Exploring and Categorizing Packages." Students classify a collection of bags, and make a chart or graph showing the number in each category.

Concepts
- Items can be grouped or classified based on characteristics they have in common.
- Data from an investigation can be displayed graphically using charts and graphs.

Vocabulary
- Category
- Classify

Skills
- Classifying objects based on material, function, size, shape, and/or other criteria
- Collecting, organizing, and presenting data in chart or graph form
- Using spoken and written language
- Problem-solving

Standards
- Benchmarks for Science Literacy: 1A, 1B, 2A, 9A
- Standards for the English Language Arts: 12
- National Science Education Standards: A
- Principles and Standards for School Mathematics: A1

Time Needed
Two to four periods

Materials
- Variety of paper and plastic bags of different colors, shapes, and sizes (e.g., paper grocery bags, other small and large paper bags, plastic grocery bags,other small and large plastic bags, assorted shopping bags, etc.)—at least 12 for each group of four students
- Chart paper
- Pencils, crayons,
- Rulers
- Worksheet #2 (at least one for each group of four students)

Procedure

1. If necessary, play the following game to review the meaning of "category" and "classify." Call a number of students to the front of the classroom who happen to be wearing the same color clothing. Ask the class what the members of this group have in common: "What's my rule for choosing these students? What is the category they all belong to?" As necessary, repeat the game using objects that have the same color, size, shape, purpose, etc.

2. Show students several types of bags and ask them to come up with ways of classifying or categorizing them. Record their answers and explanations on chart paper.

3. Divide the students into groups of four. Provide each group with at least a dozen bags and a worksheet.

4. Explain that each group should examine its bags, discuss what the bags have in common and how they are different from one another. Then they should place the bags into groups based on specific criteria.

5. The groups record their work on Worksheet #2 by listing and describing the categories, and then creating a chart to show the number of bags they placed in each category.

Tips

- Encourage students to look at the different ways the handles are made and attached to the bags.

- As an alternative or extension, have students classify groups of other packaging items such as boxes and bottles.

Worksheet #2

Classifying Bags

Name/Group _____ Date _____

How many bags do you have? _____

_____

How many groups or categories of bags did you make? _____

_____

_____

_____

Write a name for each category. _____

_____

_____

_____

Make a chart or graph showing how many bags there are in each category.

Activity Nº 3

Packing a Bag

Grade Level
Pre K-2

Prerequisites
Understanding of the concept of fair testing—using the same testing conditions for each test

Overview
This early childhood activity engages children in finding out how many blocks different types of bags can hold. They will also see the effect of the packing method on the survival of the bags. Because this activity will result in a lot of broken bags, it is a natural lead-in to bag repair activities.

Concepts
- Fair testing—a method of comparing things by testing them all in the same way.
- Design/shape, material, conditions of use, and other variables affect the strength and durability of a bag.

Vocabulary
- Fair testing
- Equal
- Same
- Different

Skills
- Measuring
- Classifying
- Problem-solving
- Drawing conclusions and making inferences based on data
- Using written and spoken language

Standards
- Benchmarks for Science Literacy: 1A. 1B, 2C
- Standards for the English Language Arts: 12
- National Science Education Standards: A
- Principles and Standards for School Mathematics: M1, DA & P3

Time Needed
Two 45-minute periods

Materials
- Two identical bags (paper or plastic) to use for fair testing demonstration
- Blocks, about 15 each of several different sizes
- Chart paper
- Sets of three or four plastic and paper bags of different sizes and styles—one set for each small group of students

Procedure: Day 1
1. Prepare ahead of time to demonstrate the concept of fair testing. Determine the number of identical blocks it takes to cause a particular paper or plastic bag to break. Use two bags of that kind, the blocks you used to cause the bag to break, and an equal number of smaller/lighter blocks for the following demonstration.

2. Label one of the identical bags A and the other B. Ask students to count with you as you place the large blocks in Bag A, one at a time. After each block, lift the bag to show that it is able to carry the weight

Keep adding blocks until the bag breaks. Write the number of blocks it took to break the bag on the front of Bag A. Then have students count with you as you add the same number of the smaller blocks to Bag B, one at a time. As before, lift the bag after each block. After you add the last block to Bag B and before you lift it, ask students, "What do you think will happen when I lift the bag?" Record their answers on chart paper. When you lift Bag B after the last block to show that the bag does not break, ask questions to encourage students to think about what happened. Record their answers on chart paper.

- "How many blocks did I put in Bag A?"

- "What happened to Bag A when I put the last block in?"

- "How many blocks did I put in Bag B?"

- "What happened to Bag B when I put the last block in?"

- "If I put the same number of blocks in both bags, why didn't Bag B break?"

- "What's the difference between Bag A and Bag B?"

- "What's the difference between the blocks I put in Bag A and the blocks I put in Bag B?"

- "If I wanted to test to see if one of these bags is stronger than the other one, what should I do differently next time?"

Record all of the students' answers to the last question.

3. Show students several bags of different sizes and/or materials. Label each bag (e.g., A, B, C, D). Ask students, "Which bag do you think will hold more blocks without breaking?" Record students' predictions and reasons on chart paper. Tally the number of students who choose each bag as the strongest.

Procedure: Day 2

1. Use your chart paper notes to review the previous day's work with students.

2. Divide students into small groups. Give each group a set of the bags you discussed and labeled on Day 1. Label the bags A, B, C, etc., as you did on Day 1.

3. Tell students their job is to test each bag to see which one holds the most blocks without breaking. Remind them that they need to use the same kinds of blocks for each bag. When a bag breaks, students count the number of blocks it took to break the bag and they write that number on the bag. (Help younger students with the counting and recording, as necessary.)

4. When a bag breaks, ask students to examine it carefully to find out what part(s) of the bag broke. Ask them to talk about why that part might have broken first.

5. Give groups time to experiment with different ways of placing the blocks in the bag or with different bags. Record the results of each experiment.

6. Follow-up with a discussion about what students discovered. Start with questions like these:

- "What did we do to figure out which bag holds the most blocks without breaking?"

- "Why did we have to use the same kinds of blocks for each bag we tested?"

- "What made the bags break?"

- "Does it matter how we put the blocks in the bag? Is there a way to put the blocks in so that the bag holds more blocks?"

- "Can you think of ways to make these bags stronger?"

Tips

- Circulate among the groups as they work so you can ask questions and hear discussions that reveal what students do and don't understand.

- Encourage students to describe and discuss their processes and ideas at every step.

- Older students might graph the results of their bag tests.

Activity № 4

It Fits Just Right!

Grade Level
Pre K-2

Prerequisites
- Knowledge and understanding of the purposes of packaging
- Ability to use problem-solving strategies

Overview
Students try to create an object that is just the right size and shape to fit in an odd-shaped package.

Concepts
- There is a relationship between the size and shape of a container and the size and shape of its contents.
- Packages can be made to fit items of a particular size and shape.

Vocabulary
- Package
- Shape
- Fit

Skills
- Applying spatial awareness to problem-solving
- Observing and comparing the physical properties of objects
- Using spoken language

Standards
- Benchmarks for Science Literacy: 1A, 1B
- Standards for the English Language Arts: 12
- Principles and Standards for School Mathematics: G1, G4

Timed Needed
One period

Materials
- Variety of containers with non-standard shapes, such as "Baci" and "Toblerone" candy boxes, "L'Eggs" pantyhose containers, partition trays for TV dinners, cookie cartons with dividers to fit the shape of the cookies, and so on
- Clay or plasticene
- Chart paper and markers

Procedure
1. Begin a discussion with the whole class by asking students if they have ever had trouble getting a book to fit in a backpack. Brainstorm examples of things that are too small, too big, or the wrong shape to fit in a package. Record students' answers and ideas on chart paper.

2. Show students some of the odd-shaped packaging materials you've collected. Ask them to imagine what kinds of objects would fit in those packages and what wouldn't fit. Ask them to explain how they can tell whether an object would fit or not.

3. Divide students into small groups. Give each group one of the packages and a lump of clay or plasticene. Ask students to make an object that they think will fit in the box or packaging partition.

4. When each group has a shape they think will fit, let them try it. If it doesn't fit, ask them to try to explain why and describe what they have to do to make it fit. Repeat this for each trial.

5. When each group has an object that fits in the packaging, have them share their processes with the other groups.

Activity № 5

How Strong Is This Bag?

Grade Level
3-6

Prerequisite
Knowledge of "fair test" principles

Overview
This is an upper-elementary-grade version of Activity #3, "Packing a Bag." Students use weights to test the strength of different types of bags. As an extension activity, students can redesign the bags to make them stronger.

Vocabulary
- Fair test
- Prediction
- Factor
- Condition

Concepts
- The strength of a bag is determined by a combination of factors, including design, materials used, and construction method.
- A "fair test" requires controlling variables so that test results can be compared.

Skills
- Collecting, organizing, and presenting data
- Making and testing predictions
- Designing and conducting an experiment, including controlling variables and analyzing results
- Measuring
- Communicating using written and spoken language

Standards
- Benchmarks for Science Literacy: 1A, 1B, 2A, 9A
- Standards for the English Language Arts: 12
- National Science Education Standards: A, B
- Principles and Standards for School Mathematics: DA & P1, DA & P3, C3, M1

Time Needed
Four to six periods

Materials
- Chart paper
- Construction paper
- Pencils and markers
- Plastic and paper bags (4 different kinds of bags for each small group of students)
- Rulers, tape measures
- Bathroom scale
- Heavy books, blocks, other objects that can be used as weights
- Worksheet #5

Procedure
1. Begin a brainstorming session with the whole class by asking questions like these:
   - Have you ever had a bag break on the way home from the store?
   - What part of the bag broke
   - Why do you think it broke?
   - What do you think could be done to prevent bags from breaking like that?"

   Encourage speculation and discussion, and record students' responses on chart paper.

2. Choose two shopping bags that are about the same size but are different in other ways—shape, material, construction. Ask students which bag would hold more weight. Encourage students to explain their reasons for thinking one bag or the other would hold more. Tell students they are going to become bag testers.

3. Review the concept of fair testing—using exactly the same conditions for every bag tested.

4. Divide the students into small groups. Provide each group with four different bags labeled A, B, C, and D. Give each group member a copy of Worksheet #5.

5. Review the worksheet and explain the assignment to students: They are to work together in teams to design a fair test that will determine which of their four bags is the strongest. Each team will:

   • examine and describe each bag;

   • predict which bag will be strongest and explain their reasons;

   • design a test that will show the relative strength of the bags;

   • carry out the test;

   • record and analyze the results;

   • organize the data in a graph or chart;

   • present the test results to the class.

6. After each group presents its test results to the class, allow time to discuss the design, the results, and possible reasons for the results. Questions like these will help get the discussion started:

   • Was this a fair test?

   • If not, how could it be changed to make it a fair test?

   • What factors make a bag weak or strong?

Extensions

• Challenge students to design tests that allow them to compare bag strength under various conditions—e.g., when the bags are wet, placing different kinds of items in the bags, using different packing techniques, and so on.

• Ask each group to examine carefully the way in which each bag failed, and to redesign their bags to make them stronger.

Tip

To read about one teacher's experience with this activity, see Chapter 4 ("Stories"), page 111.

Worksheet #5

How Strong Is This Bag?

Name/Group _____ Date _____

Bag Descriptions

Describe the four bags that you will test. Measure them and write down the measurements.
Describe their shapes, what they're made of, how they're made, and anything else you notice.

Bag #1 _____

_____

_____

Bag #2 _____

_____

_____

Bag #3 _____

_____

_____

Bag #4 _____

_____

_____

Prediction

Which bag do you think is the strongest?   ____ A   ____ B   ____ C   ____ D

Why do you think that one is the strongest? _____

_____

Test

Describe the test you will use to find out which bag is the strongest _____

_____

Test Results

Based on your test, which bag is the strongest? _____

_____

Activity № 6

How Do You Package a Fragile Object?

Grade Level
4-6

Prerequisites
- Understanding of packaging materials
- Understanding of categorizing
- Understanding of fair test principles

Overview
Students test different kinds of cushioning materials and different ways of arranging them in order to protect the fragile contents of a package.

Concepts
- Packaging is designed to meet different needs and solve different problems.
- Comparing and evaluating several solutions to the same problem requires the use of the principles of fair testing.
- The composition and arrangement of different materials affect their ability to protect fragile objects in packages.

Vocabulary
- Fair test
- Fragile
- Cushioning

Skills
- Analyzing and solving problems
- Reasoning
- Using written and spoken language
- Collecting, organizing, and interpreting data
- Using charts and graphs to present data

Standards
- Benchmarks for Science Literacy: 1B, 3B, 8B
- National Science Education Standards: A, E
- Standards for the English Language Arts: 12
- Principles and Standards for School Mathematics: DA & P1, DA & P3, C3, M1,

Time Needed
Four to six periods

Materials
- Variety of cushioning materials such as foam rubber, Styrofoam sheets and "peanuts," newspaper, bubble wrap, cloth, towels, cotton
- Fragile objects such as chalk, thin bread sticks, cookies
- Variety of unbreakable packaging containers (small boxes, milk containers, padded envelopes)
- Measuring tape
- Tape
- Paper
- Pencils and markers
- Worksheet #6

Procedure
1. Place some fragile objects—cookies, breadsticks, pieces of chalk—in a small, clear plastic bag and seal the bag. Ask students what they think will happen to the contents if you drop the bag. Once students have offered their ideas, drop the bag on a hard surface from a height of at least three feet. Show the package to the students again so they can see that the contents are broken.

2. Ask students to brainstorm ways the same items could be packaged so they wouldn't break when the package was dropped. Record their responses on chart paper.

3. Divide students into small groups. Explain that their job is to test different ways of packaging fragile objects in order to determine which method does the best job of protecting the objects when the package is dropped.

4. Review the rules of fair testing. Remind students that each test must be conducted under the same conditions—in this case, each package must contain the same type of fragile object and must be dropped from the same height.

5. Provide all groups with at least three different packaging containers, several different kinds of cushioning materials, fragile items (e.g., cookies, breadsticks, pieces of chalk), and a copy of Worksheet #6. Go over the worksheet with the class to make sure everyone understands what to do. Help groups figure out how to make sure they drop all packages from

the same height—e.g., pushing each package gently off a table or a shelf.

6. Allow two or three class periods for groups to set up, conduct, and record the results of their tests. Observe the groups as they work and help them record each test accurately and completely—kind of container, kind and amount of cushioning material, condition of the fragile item after each test.

7. All presentations should be visual, including demonstrations and references to charts and/or graphs.

Tips

For one teacher's experience with this activity, see Chapter 4, "Stories," page 108. An egg is often used in this kind of activity, but cookies, breadsticks, or chalk are less messy and also less wasteful, since broken cookies and breadsticks can still be eaten, and broken chalk can still be used.

Worksheet #6

Testing Fragile Objects

Name/Group _____   Date _____

Packaging Used	Height	Type of Material (cushion)	Condition of Fragile Object

Activity №7

Which Pump Dispenser Works Best?

Grade Level
5–6

Prerequisites
- Understanding of "fair testing" principles
- Knowledge of experimental design, including isolating variables

Overview
This is a product testing activity, in which students formulate their own criteria for the "best" pump. Then they conduct systematic tests to find out how well a variety of pumps meet the criteria. They also try the pumps on different liquids, to see whether the best pump for a fluid like water is also most effective with a viscous fluid such as ketchup.

Concepts
- Different kinds of fluids have different viscosities
- Variables that can affect the functioning of a pump or spray dispenser include the viscosity of the liquid and the design of the device

Vocabulary
- Viscosity
- Pump
- Dispenser
- Stroke

Skills
- Using problem-solving strategies to design an experiment
- Measuring
- Creating and collecting data
- Analyzing data and drawing conclusions from evidence
- Representing data on charts or graphs

Standards
- Benchmarks for Science Literacy: 1A, 1B, 9A
- Standards for the English Language Arts: 12
- National Science Education Standards: A
- Principles and Standards for School Mathematics: DA & P1, DA & P3, C3, M1,

Time Needed
10 to 14 periods

Materials
- Pump and spray dispensers from the tops of containers for water, soap, lotion, mustard, ketchup, liquid cleaners, and so on— one push pump and one spray dispenser for each small group of students
- Liquids of different viscosities, such as water, milk, liquid detergent, ketchup, mustard
- Basins or buckets
- Plastic cups
- Newspaper, paper towels
- Construction paper
- Colored pencils, markers
- Meter sticks or yardsticks
- Measuring cups or graduated cylinders
- Rulers
- Chart paper
- Worksheets #7A, #7B, #7C

Procedure
1. Divide the students into groups. Give each group two pumps—one spray pump used for water or other thin liquids and one pump used for thick liquids such as liquid hand soap or dishwashing detergent, mustard, or ketchup; a ruler;

paper; and colored pencils. Ask students to examine the pumps and liquids carefully, draw the pumps, and write down their observations, focusing on these questions:

- How is the spray pump different from the push pump?
- How does each pump work?
- What is each pump used for?
- How is water different from liquid detergent, mustard, or ketchup?

2. Bring the groups together and have students share their observations with the whole class. Continue with a brainstorming session about the question, "How could we figure out which kind of pump works best for different kinds of liquids—water, milk, liquid detergent, ketchup?" Help students come up with several testing strategies—e.g., measuring how much liquid comes out with each pump; counting the number of times the pump has to be pressed or squeezed (i.e., the number of strokes) in order to remove a certain amount of liquid from a container; and so on. Record all of the students' ideas on chart paper.

3. Distribute copies of Worksheets #7A, 7B, and 7C to all students. Explain that students will work in their groups to test the pumps to determine which pump works best for different kinds of liquids. In their groups, students discuss methods for testing the pumps for different liquids. Each group should agree on a method for testing the two kinds of pumps with different kinds of liquids. Have each student complete Worksheet #7A.

4. Give each group three or four liquids to test with the two kinds of pumps. (See Tips, below.) Have an assortment of plastic measuring cups, graduated cylinders, and measuring spoons available. Allow one or more class periods for students to design, carry out, and document their tests, and analyze their data.

5. Each group prepares their test results to present to the class in the form of a chart or graph. Set aside class time for group presentations.

6. Follow-up student presentations with a discussion of the findings, using questions like these to get the discussion going:

- Does the size of the tube affect the way a pump works for a particular kind of liquid?
- Would making the tube bigger make it work better for thick liquids?
- What are some other ways packages are designed to help you get the contents out of the container?

Tips

- Students need to test several different kinds of liquids with each pump. Try to provide each group with two relatively thin liquids, such as water and milk, and two relatively thicker liquids, such as liquid detergent and ketchup.

- Avoid using lotions and oils for this activity, since they make the pumps difficult to clean and reuse.

- This is a messy activity. Cover students' workspaces with newspaper, and have plenty of paper towels and sponges available for cleanup.

Worksheet #7A

Pump Test Predictions and Procedures

Name/Group _____ Date _____

Problem:

What are you trying to find out?

_____

_____

Hypothesis:

List the liquids you will test. Then predict which kind of pump (spray or push) will work best for each liquid.

Liquid: _____

Prediction: _____

Liquid: _____

Prediction: _____

Liquid: _____

Prediction: _____

Liquid: _____

Prediction: _____

Procedure:

How will you test the pumps? What steps will you follow?

_____

_____

_____

_____

Worksheet #7B

Pump Test Data

Name/Group   Date
_____

Create a chart or data table to record the data you collected.

Worksheet #7C

Pump Test Findings and Analysis

Name/Group _____ Date _____

List each liquid you tested and the pump that worked best for each one.

Liquid: _____

Pump that worked best for this liquid: _____

Why do you think this pump worked better for this liquid? _____

Liquid: _____

Pump that worked best for this liquid: _____

Why do you think this pump worked better for this liquid? _____

Liquid: _____

Pump that worked best for this liquid: _____

Why do you think this pump worked better for this liquid? _____

Liquid: _____

Pump that worked best for this liquid: _____

Why do you think this pump worked better for this liquid? _____

What else did you learn from doing this experiment? _____

_____

_____

_____

Part Two

Cardboard Structures

Overview

This unit consists of six activities that look systematically at some of the variables involved in constructing cardboard structures. Each activity examines a different variable through a controlled experiment. The variables are:

• the shape of a shelf or beam;

• the type of cardboard used;

• the orientation of the cardboard "ribs";

• the type of glue used to laminate sheets of cardboard together; and

• the method used to support the shelves or beams.

In addition, the first activity also teaches the principles of a controlled experiment or "fair test."

Some of the variables studied in this unit could be very important in the design of a usable classroom shelving unit, display stand, chair, or table. Each activity could also be the basis for a science fair project.

Prerequisites

• Knowledge of the principles of a "fair test"

• Experience in measuring weight and length

• Knowledge of geometric shapes

Vocabulary

• Structure

• Force

• Load

• Stability

• Tension

• Compression

• Beam

• Column

• Balance

• Center-of-mass

• Lamination

• Support

Concepts

• The strength and point at which a structure fails depend on variables including the shape of beams and columns, type and use of materials, construction methods

• Different parts of a structure are subject to different forces

Skills

• Identifying variables

• Collecting and organizing data

• Using problem-solving strategies in mathematics and language arts

• Measuring

• Drawing conclusions

• Presenting results in written and spoken language

Standards

• Benchmarks for Science Literacy: 1A, 1B, 3B, 8B, 12A

• National Science Education Standards: A, B, E

• Standards for the English Language Arts: 12

• Principles and Standards for School Mathematics: DA & P1, DA & P3, C3, M1, G1

Grade Level

5-6

Time Needed

3 to 5 periods for each activity

Activity Sequence:

#8: How Does the Shape of a Column Affect Its Strength?

#9: How Does the Shape of a Shelf Affect Its Strength?

#10: How Does the Type of Cardboard Affect the Strength of a Shelf?

#11: How Does the Direction of the Corrugations Affect the Strength of a shelf?

#12: How Does the Type of Glue Affect the Strength of a Laminated Shelf?

#13: How Does the Support Method Affect the Strength of a Shelving Unit?

Activity № 8

How Does the Shape of a Column Affect Its Strength.

Overview

In this activity, students investigate the relative strength of columns of different shapes and discover the importance of controlling variables in a test or experiment.

Materials

- Paper of assorted sizes and shapes (for preliminary construction)
- Shape templates (pages 69-72)
- Tape or glue
- Weights for load testing (tiles, washers, or marbles)
- Bucket or coffee can
- Square pieces of cardboard large enough to support a can or bucket
- Worksheet #8

Procedure

1. A column is a vertical piece whose purpose is to support parts of a structure, such as a building, a platform, or a bookshelf. The legs of tables and chairs also work as columns. Find a structure in the classroom or nearby in the school that has support columns. This could be part of the room or building, a raised platform, a large table, or a bookshelf. Ask students to observe the columns carefully, and answer questions like these:

- What purpose do these columns serve in this structure? What are they holding up?
- What would happen if one or more of the columns were removed?
- What would happen if the columns weren't strong enough to do their job of supporting the structure?
- What makes a column strong? Is it the length, the diameter, the material, the shape, or a combination of those?
- How would we measure the length of a column?
- How would we measure the circumference or perimeter of a column?
- What kinds of materials can columns be made of?
- What shapes can columns be?

Regarding the latter question, show columns that are in the shape of a circle, a square, a rectangle, and a triangle. You can find some of these in furniture or in the structure of a building, or make them out of paper or cardboard for demonstration purposes.

Record students' answers so they can return to them later after they've done some tests to learn more about this topic.

2. Draw four shapes on the chalkboard: square, triangle, rectangle, and circle. Take a vote among students on which of these shapes they think would make the strongest column. Ask them to give their reasons for their choices. Don't label answers as right or wrong. Instead, record their ideas and tally up the votes for each shape for future reference.

3. Ask students to work in pairs to construct the four different-shaped columns out of paper and use them to test their hypotheses. Don't specify any rules or guidelines for the sizes of the columns or the construction techniques. Make paper of different weights and sizes available, but don't suggest which one should be used. Let students make all the decisions regarding the columns. The only requirement is that one must be round, one must be square, one must be rectangular, and one must be triangular.

4. Distribute copies of Worksheet #8 to each team. On the worksheet, have them identify this as "Test #1." Once students have constructed their columns, have them measure the length and enter the measurements on Worksheet #8. They should also enter the construction material used for each column.

5. The next step is for students to test the columns for strength and record the results. The test consists of centering a square platform of cardboard on top of the column, holding the column securely and making sure it is exactly vertical, placing the can on top of this platform, and adding weights (identical washers, tiles, or marbles) to the can until the column buckles. One student can hold the column upright at its base while the other places the cardboard platform on top, places the empty can on the platform, then adds the weights to the can, one at a time. Students should count the number of weights added in each test to determine the point at which the column buckles.

6. Make a simple chart on the board or chart paper to compare the results of all student teams. Tally how many columns of each shape were found to be the strongest in the preliminary tests. If a clear winner emerges, ask student teams who identified that shape as the strongest to share their data with the class:

   • Length of column

   • Material used

   • Point of failure

   Ask students to compare and discuss similarities and differences among these variables.

7. Discuss the meaning of these preliminary tests, starting with these questions:

   • Why didn't every team get the same result?

   • Do these tests really reveal which shape makes the strongest column?

   • Were these fair tests—that is, were the variables (such as length, width, type of paper) the same for each test?

8. Review the principles of fair testing with students. Have them identify what would be necessary to make these fair tests for the strongest shape— that is, controlling all variables so that the only difference among the columns being tested is the shape. This requires using the same materials, making columns of the same length, and using the same platform, can, and weights to test for strength.

9. Once the principles of a fair test have been established, set aside class time for students to perform their tests again. Distribute more copies of Worksheet #8 to student teams and have them identify this as "Test #2." Make copies of the templates on pages 69-72 and distribute them to students. Point out that it's important to follow the template directions carefully so that all columns are folded and taped in the same way.

10. Have students make the columns and perform the tests again. Remind them to follow the principles of fair testing in all cases—all variables must be the same except the shape of the column.

11. Students once again fill out the worksheet and present their findings to the class. Compare and discuss the findings, which should reveal the round column to be the strongest. If there are results that don't agree with this finding, have students analyze their test procedures and conditions to look for evidence that fair test guidelines were not followed.

Extensions

• Challenge students to create a graph of their test data and to rank the four shapes from strongest to weakest.

• As homework, have students look for examples of columns they see in their homes and community, and also find pictures of columns in newspapers and magazines and bring them to class. Set aside class time to discuss the shapes, materials, and functions of the columns students identify.

Worksheet #8

Which Shape Makes the Strongest Column?

Name/Team _____   Date _____

Test #_____

Shape of Column	Length of Column	Construction Material	Number of Weights That Caused Failure
Circle			
Square			
Rectangle			
Triangle			

1. Which column was the strongest?   ____ Circle   ____ Square   ____ Rectangle   ____ Triangle

2. Why do you think that one is the strongest? _____

_____

_____

3. Was it the same length as the other columns?   ____ Yes   ____ No

4. Was it made of the same material as the other columns?   ____ Yes   ____ No

5. How do you think the answers to questions 3 and 4 affected the test results? _____

_____

_____

_____

_____

(tape here)

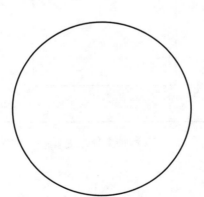

To make a circular column, tape this edge of paper to the line at the other end.

      ↓   ↓   ↓   ↓   ↓

(fold here and tape)

(fold here)

(fold here)

(fold here)

To make a square column, fold at the lines and tape this edge of paper to the line at the other end of the paper.

↓　　　↓　　　↓　　　↓　　　↓　　　↓　　　↓　　　↓

(TAPE HERE but DO NOT FOLD AT THIS LINE!)

(fold here)

(fold here)

(fold here)

(fold here)

To make a rectangular column, fold at the lines and tape this edge of paper to the line at the other end of the paper.

↓   ↓   ↓   ↓   ↓   ↓   ↓   ↓

(fold here and tape)

(fold here)

(fold here)

To make a triangular column, fold at the lines and tape this edge of paper to the line at the other end of the paper.

↓   ↓   ↓   ↓   ↓   ↓   ↓   ↓

Activity №9

How Does the Shape of a Shelf Affect Its Strength?

Overview

Students conduct tests to determine which shape will produce the strongest shelf.

Materials

- Cardboard (to make square, rectangular, and trapezoidal shelves)
- Strong, wide cardboard columns (two per group) that the shelves will be attached to for testing
- Glue
- Heavy weights for load testing
- Meter stick
- Worksheet #9

Procedure

1. Explain that the goal of this activity is to find the shape that makes the strongest shelf: square, rectangle, or trapezoid. Students will work in teams to test the shapes and compare their results.

2. Divide students into small groups. Assign a shape to each group. At least two groups should test each shape so that results can be compared.

3. Provide groups with materials for making shelves—cardboard for the shelves, cardboard support columns, glue or tape for attaching shelves to columns. If students are cutting their own cardboard, provide them with dimensions so that all shelves tested have the same surface area—required to make this a fair test.

4. Review methods for testing shelf strength. There are two options:

   - *Drop method:* Using the meter stick as a guide, students drop a standard weight onto the center of the shelf from 5 cm. above it. They repeat this test, dropping the same standard weight from heights that increase in 5 cm. increments, until the shelf gives way from the impact.

   - *Gradual method:* Students load the shelf by gradually adding weight to the center of the shelf, until it fails.

5. Distribute Worksheet #9. Have students conduct the tests and record their findings in the appropriate place, depending on the test method used.

6. Bring the groups together and ask each group to share its results. Create a chart on the board or chart paper to show the results for each shape from each group. Compare and discuss the results, starting with these questions:

   - Which shape was the strongest?

   - Did each group have similar results?

   - If different loading methods were used with the same shape, do the results from both types of test agree?

Encourage discussion on whether the "drop" method or the "gradual" method better reflects the way a shelf would be loaded in practice.

Tips

• Monitor the groups' work to make sure students glue or tape the shelves securely to the columns. If glue is used, it should be allowed to dry overnight. If the shelves aren't attached securely to the columns or if the glue isn't dry, the supports could fail before the shelves do.

• The surface area should be the same for each shape. You can assure this by giving students the dimensions for cutting the shelves, or by precutting the cardboard pieces yourself. Alternatively, the students can calculate the dimensions themselves, as a math exercise. Other variables that should be kept the same include the type of cardboard and the support method.

• The simplest approach is to have each group test a different shape and then compare results. A more thorough but time-consuming approach is to allow each group to test all three shapes.

• In order for this to be a fair test, there are a number of variables that must be kept constant, such as the surface area of the shelves. Review the concepts of a controlled experiment.

• If you use the "gradual" method of testing, you may need to make the shelf longer so that it requires a reasonable amount of weight to make it fail.

Worksheet #9

How Does the Shape of a Shelf Affect Its Strength?

Name/Group   Date

Test Method: Drop

Shape of Shelf	Dropped 5 cm.	Dropped 10 cm.	Dropped 15 cm.	Dropped 20 cm.

Test Method: Gradual

Fill in the shelf shape beng tested. Enter the amount of weight loaded onto the shelf that caused the shelf to fail.

Shape of Shelf	Number of Weights at Which the Shelf Failed

Activity № 10

How Does the Type of Cardboard Affect the Strength of a Shelf?

Overview

Students conduct tests to determine which type of cardboard produces the strongest beam or shelf.

Materials

- Three or four common types of cardboard of different weights and thicknesses, large enough to make rectangular beams approximately 6 cm. x 30 cm.

- Weights for load testing (washers, tiles, or marbles)

- Bucket or can

- Worksheet #10

Procedure

1. Explain that the goal of this activity is to find out which type of cardboard makes the strongest shelf. Students will work in small groups to test different kinds of cardboard.

2. Divide students into small groups. Assign a type of cardboard to each group. At least two groups should test each type of cardboard so that results can be compared.

3. Provide groups with materials for making the rectangular beams (shelves). If students are cutting their own cardboard, provide them with dimensions—6 cm. x 30 cm.— so that all cardboard beams tested have the same surface area—required to make this a fair test. Remind students of the principles of the fair test and review the variables in this test that must be kept constant.

4. Have groups test each piece of cardboard using this procedure:

   - Use the cardboard rectangle to make a "bridge" between two desks or two stacks of books.

   - Place the bucket or can at the center of the cardboard.

   - Gradually add tiles or washers (or other appropriate weights) to the bucket, until the shelf collapses. Keep track of the number of weights added.

   - Record the number of weights that cause the cardboard "bridge" to fail on Worksheet #10.

5. Bring the groups together to report, compare, and discuss the test results.

Tips

- You will probably want to limit the types of cardboard to three or four types that are large enough to work with and readily available.

- To save time and to insure that all cardboard rectangles are the same size, you may want to pre-cut the cardboard pieces yourself. If you have students cut their own cardboard, monitor them carefully to make sure they are using cutting tools safely and to make sure the pieces tested are all the same size.

- The simplest approach would be to have each group test one or two different materials, then compare results. A more thorough but time-consuming approach would be to allow each group to test all three or four materials.

Worksheet #10

How Does the Type of Cardboard Affect the Strength of a Shelf?

Name/Group _____ Date _____

Type of cardboard tested:

_____

Dimensions of cardboard shelf:

_____

Kind of weights used to test strength:

_____

Number of weights that caused the shelf to fail:

_____

Activity №11

How Does the Direction of the Corrugations Affect the Strength of a Shelf?

Overview

Students conduct tests to determine which way of orienting the corrugations will produce the strongest shelf.

Materials

- Three 6 cm. x 30 cm. pieces of cardboard per group
- Glue
- Tiles, washers, or marbles for the load testing
- Bucket or large can for holding weights
- Worksheet #11

Procedure

1. Provide students with square samples of corrugated cardboard and let them examine them carefully. Discuss the structure of corrugated cardboard. Help students observe that the middle layer has "ribs" that all run in the same direction. Ask these questions and record students' answers. Encourage students to speculate and give reasons for their answers.

   - Would a cardboard shelf be stronger if these ribs were running the long way or the short way?

   - If three pieces of cardboard are glued together to make a stronger shelf, should all of the pieces have their ribs running the same way, or should they be alternated? If they should alternate, what should the order be?

2. Divide students into small groups. Provide each group with two 6 cm. x 30 cm. "shelves" made of the same kind of corrugated cardboard. In one shelf, the ribs of the cardboard run from end to end of the shelf—the long way; in the other, the ribs run across the shelf—the short way.

3. Distribute copies of Worksheet #11. Have all groups test the strength of each shelf following this procedure:

   - Use the cardboard shelf to make a "bridge" between two desks or two stacks of books.

   - Place the bucket or can at the center of the cardboard.

   - Gradually add tiles or washers (or other appropriate weights) to the bucket, until the shelf collapses. Keep track of the number of weights added.

   - Record the number of weights that cause the cardboard "bridge" to fail.

4. Bring the groups together to present, compare, and discuss the results. They should discover that cardboard is stronger when its ribs run the long way.

5. Pose the next question: If three pieces of cardboard are glued together to make a stronger shelf, should all of the pieces have their ribs running the same way, or should they be alternated? If they should alternate, what should the order be? Record students' responses. Encourage them to speculate and to explain the reasons for their answers.

6. Explain to students that they are going to design and test a model of a laminated shelf made from three 6 cm x 30 cm pieces of corrugated cardboard. A laminated shelf is assembled by gluing three layers of shelving material together.

7. Have each group choose one of the laminated shelf patterns listed under Test #2 on Worksheet #11. "Long/long/long" means all three layers have their ribs running the long way, etc. Make sure that at least one group is testing each design. Have all groups make and test their shelf designs using this procedure:

- Choose a shelf design and check it off on Worksheet #11.

- Create the laminated shelf by gluing three pieces of 6 cm. x 30 cm. corrugated cardboard together.

- Use a book as a weight to hold the pieces together and allow the assembled shelf to dry, preferably overnight.

- Use the laminated shelf to make a "bridge" between two desks.

- Place the bucket or can at the midpoint of the shelf.

- Add weights, one at a time, until the shelf collapses.

- On Worksheet #11, record the number of weights that caused the collapse.

8. Bring the groups together to present, compare, and discuss the test results. The results should show that the pattern long/long/long makes the strongest shelf.

Tips

- To save time, you may want to prepare the cardboard ahead of time, or you can allow students to cut their own cardboard. All shelf pieces must be the same size and made of the same type of corrugated cardboard. You need enough pieces with the ribs running the long way, and enough oriented the short way, so that all variations can be tested.

- Review the principles of fair testing. To insure that this is a fair test, all teams should use the same amount of glue to create their laminated shelves, the same types of buckets or cans, and the same types of weights.

Worksheet #11

How Does the Direction of the Corrugations Affect the Strength of a Shelf?

Name/Group   Date

Test #1: Single-Layer Shelf

Shelf Description	Number of Weights at Which Shelf Failed
Ribs running the long way	
Ribs running the short way	

Test #2: Three-Layer Laminated Shelf

Shelf Design Tested (check one)	Number of Weights at Which Shelf Failed
[ ] Long/long/long	
[ ] Long/long/short	
[ ] Long/short/long	
[ ] Long/short/short	
[ ] Short/long/short	
[ ] Short/short/short	

Activity № 12

How Does the Type of Glue Affect the Strength of a Laminated Shelf?

Overview
Students conduct tests to determine which glue will produce the strongest shelf.

Materials
- Glue of different types, such as wood glue, white school glue, mucilage, stick glue, rubber cement, art glue, fabric glue, etc.
- Three 6 cm. x 30 cm. pieces of cardboard per group
- Tiles, washers, or marbles for load testing
- Bucket or large can for holding weights
- Worksheet #12

Procedure
1. Explain to students that the goal of this activity is to find the glue that will make the strongest laminated shelves. Students will work in groups to test the strength of different glues used to create a two-layer laminated shelf.

2. Divide students into small groups. Provide each group with two identical 6 cm. x 30 cm. cardboard shelves and a standard amount of one type of glue. In this test, the directions of the cardboard ribs, the size, and type of cardboard must all be the same, as must the amount of glue used.

3. Distribute copies of Worksheet #12. Have all groups conduct a test on one type of glue using the following procedure:
   - On Worksheet #12 check the type of glue being tested.
   - Carefully use all the glue provided to glue the two cardboard layers together.
   - Put weights on the shelves and allow them to dry, preferably overnight.
   - When the glue is dry, use the laminated shelf to make a bridge between two desks or two stacks of books.
   - Place a can or bucket in the center of the shelf and add tiles, washers, or other weights, one at a time, until the shelf collapses.
   - On Worksheet #12 record the number of weights at which the shelf collapses.

4. Bring the groups together to present, compare, and discuss the results. Use these questions to spark discussion:
   - Would we get the same results with a different type of cardboard?
   - Would the same glue still be strongest if it were used for a different purpose, such as making a column or supporting a shelf?

Tips
- At least two groups should test each type of glue, so the results can be compared.
- To save time and insure that this is a fair test, you might want to prepare the cardboard layers ahead of time.
- Pre-measure the same amount of glue for each group.
- When using rubber cement or any other glue with a strong odor, make sure there is plenty of ventilation.

Worksheet #12

How Does the Type of Glue Affect the Strength of a Laminated Shelf?

Name/Group   Date

Type of Glue Tested (check one)	Number of Weights at Which Shelf Failed
[ ] Wood glue	
[ ] White school glue	
[ ] Mucilage	
[ ] Stick glue	
[ ] Rubber cement	
[ ] Art glue	
[ ] Fabric glue	
[ ] Other (describe)	

Activity № 13

How Does the Support Method Affect the Strength of a Shelving Unit?

Overview

Students conduct tests to determine which type of support will produce the strongest shelf unit.

Materials

- Cardboard to build one model unit for each group (sides and back only)
- Shelves from previous activities
- Three narrow strips of cardboard to serve as supports (two for the sides and one for the back)
- White school glue
- Tape
- Tiles, washers, or marbles to use as weights for load testing
- Bucket or large can for holding weights
- Worksheet #13

Procedure

1. Explain to students that they are going to test different methods for supporting the shelves of a cardboard shelving unit to figure out which method works best. Students will work in groups to choose and test different methods.

2. Brainstorm ways that shelves might be supported in a shelving unit. Have students look around the room to see how the shelves are supported in bookcases, closets, storage cabinets, etc. Record all of the students' ideas on the board or chart paper. If students need help, suggest these methods:

   - Strips of cardboard are glued to sides and back of unit and the shelves then rest on these strips.
   - Shelves are glued directly to the backs and sides of the unit.
   - Tape is used to hold the shelves directly to sides and back of unit.

3. Divide students into small groups and distribute copies of Worksheet #13. Let each group choose the attachment method they will test. Make the necessary materials available.

4. Give groups time to attach shelves to the shelving unit frame you have provided. When the unit is complete, the group uses the can or bucket with weights to test the unit for strength. They then record their results on Worksheet #13.

5. Bring the groups together to present, compare, and discuss the results.

Tips

- For this activity you will need to prepare a model unit for each group consisting of the back and sides only.

- In order for this to be a fair test, variables must be kept the same for all trials: the dimensions and materials used in the frame; the size of the shelves and materials used.

Worksheet #13

How Does the Support Method Affect the Strength of a Shelving Unit?

Name/Group _____   Date _____

Describe the method you will use to attach the shelves to the shelving unit.

_____

_____

_____

_____

Describe what happened when you tested the strength of your shelving unit.

_____

_____

_____

_____

How could you improve the design of your shelving unit?

_____

_____

_____

_____

Standards for Activities

Activity #1: Exploring and Categorizing Packages

Benchmarks for Science Literacy
Benchmark #1B: Describing things as accurately as possible is important in science because it enables people to compare their observations with those of others.

Standards for the English Language Arts
Standard #12: Students use spoken written language to accomplish their own purposes.

National Science Education Standards
Content Standard A: Students should develop abilities necessary to do scientific inquiry.

Principles and Standards for School Mathematics
Algebra Standard A1: Understand, patterns, relations, and functions.

Activity #2: Classifying Bags

Benchmarks for Science Literacy
Benchmark #1A: Results of similar scientific investigations seldom turn out exactly the same. Sometimes this is because of unexpected differences in the things being investigated.
Benchmark #1B: Scientific investigations may take many different forms including observing what things are like or what's happening somewhere, collecting specimens for analysis, and doing experiments.
Benchmark #2A: Mathematics is the study of many kinds of patterns, including numbers and shapes and operations on them. Sometimes patterns are studied because they help to explain how the world works or how to solve practical problems.
Benchmark #9A: Simple graphs can help to tell about observations.

Standards for the English Language Arts
Standard #12: Students use spoken, written, and visual language to accomplish their own purposes.

National Science Education Standards
Content Standard A: Students should develop abilities necessary to do scientific inquiry.

Principles and Standards for School Mathematics
Algebra Standard A1: Understand, patterns, relations, and functions
Data Analysis and Probability Standard DA & P1: Formulate questions that can be addressed with data and collect, organize, and display relevant data to answer them.

Activity #3: Packing a Bag

Benchmarks for Science Literacy

Benchmark #1A: When a scientific investigation is done the way it was done before, we expect to get a very similar result.

Benchmark #1B: People can often learn about things around them by just observing those things carefully, but sometimes they can learn more by doing something to the things and noting what happens.

Benchmark #2C: Numbers and shapes can be used to tell about things.

Standards for the English Language Arts

Standard #12: Students use spoken, written, and visual language to accomplish their own purposes.

National Science Education Standards

Content Standard A: Students should develop abilities necessary to do scientific inquiry.

Principles and Standards for School Mathematics

Measurement Standard M1: Understand measurable attributes of objects and the units, systems, and processes of measurement.

Data Analysis and Probability Standard DA & P3: Develop and evaluate inferences and predictions that are based on data.

Activity #4: It Fits Just Right!

Benchmarks for Science Literacy

Benchmark #1A: When a scientific investigation is done the way it was done before, we expect to get a very similar result.

Benchmark #1B: People can often learn about things around them by just observing those things carefully, but sometimes they can learn more by doing something to the things and noting what happens.

Standards for the English Language Arts

Standard #12: Students use spoken, written, and visual language to accomplish their own purposes.

Principles and Standards for School Mathematics

Geometry Standard G1: Analyze characteristics and properties of two- and three-dimensional geometric shapes and develop mathematical arguments about geometric relationships.

Geometry Standard G4: Use visualization, spatial reasoning, and geometric modeling to solve problems.

Activity #5: How Strong Is This Bag?

Benchmarks for Science Literacy

Benchmark #1A: When a scientific investigation is done the way it was done before, we expect to get a very similar result.

Benchmark #1B: People can often learn about things around them by just observing those things carefully, but sometimes they can learn more by doing something to the things and noting what happens.

Benchmark #2A: Mathematics is the study of many kinds of patterns, including numbers and shapes and operations on them. Sometimes patterns are studied because they help to explain how the world works or how to solve practical problems.

Benchmark #9A: Simple graphs can help to tell about observations.

Standards for the English Language Arts

Standard #12: Students use spoken, written, and visual language to accomplish their own purposes.

National Science Education Standards

Content Standard A: Students should develop abilities necessary to do scientific inquiry.

Content Standard B: Students should develop an understanding of properties of objects and materials.

Principles and Standards for School Mathematics

Data Analysis and Probability Standard DA & P1: Formulate questions that can be addressed with data and collect, organize, and display relevant data to answer them.

Data Analysis and Probability Standard DA & P3: Develop and evaluate inferences and predictions that are based on data.

Connections Standard C3: Recognize and apply mathematics in contexts outside of mathematics.

Measurement Standard M1: Understand measurable attributes of objects and the units, systems, and processes of measurement.

Activity #6: How Do You Package a Fragile Object?

Benchmarks for Science Literacy

Benchmark #1B: Describing things as accurately as possible is important in science because it enables people to compare their observations with those of others.

Benchmark #3B: There is no perfect design.

Benchmark #8B: Discarded products contribute to the problem of waste disposal. Sometimes it is possible to use the materials in them to make new products, but materials differ widely in the ease with which they can be recycled.

National Science Education Standards
Content Standard A: Students should develop abilities necessary to do scientific inquiry.
Content Standard E: Students should develop abilities of technological design and understanding about science and technology.

Standards for the English Language Arts
Standard #12: Students use spoken, written, and visual language to accomplish their own purposes.

Principles and Standards for School Mathematics
Data Analysis and Probability Standard DA & P1: Formulate questions that can be addressed with data and collect, organize, and display relevant data to answer them.
Data Analysis and Probability Standard DA & P3: Develop and evaluate inferences and predictions that are based on data.
Connections Standard C3: Recognize and apply mathematics in contexts outside of mathematics.
Measurement Standard M1: Understand measurable attributes of objects and the units, systems, and processes of measurement.

Activity #7: Which Pump Dispenser Works Best?

Benchmarks for Science Literacy
Benchmark #1A: When a scientific investigation is done the way it was done before, we expect to get a very similar result.
Benchmark #1B: Describing things as accurately as possible is important in science because it enables people to compare their observations with those of others.
Benchmark #9A: Simple graphs can help to tell about observations.

Standards for the English Language Arts
Standard #12: Students use spoken, written, and visual language to accomplish their own purposes.

National Science Education Standards
Content Standard A: Students should develop abilities necessary to do scientific inquiry.

Principles and Standards for School Mathematics
Data Analysis and Probability Standard DA & P1: Formulate questions that can be addressed with data and collect, organize, and display relevant data to answer them.
Data Analysis and Probability Standard DA & P3: Develop and evaluate inferences and predictions that are based on data.
Connections Standard C3: Recognize and apply mathematics in contexts outside of mathematics.
Measurement Standard M1: Understand measurable attributes of objects and the units, systems, and processes of measurement.

Activities #8-13: Cardboard Structures

Benchmarks for Science Literacy

Benchmark #1A: When a scientific investigation is done the way it was done before, we expect to get a very similar result.

Benchmark #1B: Describing things as accurately as possible is important in science because it enables people to compare their observations with those of others.

Benchmark #3B: There is no perfect design.

Benchmark #8B: Discarded products contribute to the problem of waste disposal. Sometimes it is possible to use the materials in them to make new products, but materials differ widely in the ease with which they can be recycled.

Benchmark #12A: Students should keep records of their investigations and observations and not change the records later. Students should offer reasons for their findings and consider reasons suggested by others.

National Science Education Standards

Content Standard A: Students should develop abilities necessary to do scientific inquiry.

Content Standard B: Students should develop an understanding of properties of objects and materials.

Content Standard E: Students should develop abilities of technological design and understanding about science and technology.

Standards for the English Language Arts

Standard #12: Students use spoken, written, and visual language to accomplish their own purposes.

Principles and Standards for School Mathematics

Data Analysis and Probability Standard DA & P1: Formulate questions that can be addressed with data and collect, organize, and display relevant data to answer them.

Data Analysis and Probability Standard DA & P3: Develop and evaluate inferences and predictions that are based on data.

Connections Standard C3: Recognize and apply mathematics in contexts outside of mathematics.

Measurement Standard M1: Understand measurable attributes of objects and the units, systems, and processes of measurement.

Geometry Standard G1: Analyze characteristics and properties of two- and three-dimensional geometric shapes and develop mathematical arguments about geometric relationships.

Chapter 4

STORIES

in this chapter, seven teachers will tell the stories of how they implemented activities and units on packaging and structures at levels ranging from pre-kindergarten/kindergarten to sixth grade. Five of the teachers focused on analysis and design of packaging technologies. The other two engaged their students in designing and making useful classroom structures from discarded packaging materials. All of the activities in this chapter are summarized in Table 4-1.

Table 4-1
Approaches to the study of packaging and structures

Getting Started

Brianstorming about what constitutes a package, including natural and artificial packages

Collecting and sorting examples of packaging

Brainstorming problems at school and home that could be solved by creating new structures

Analysis

Exploring some of the relationships between a package and its contents

Testing bags, boxes, cushioning materials, and pump dispensers to find out their mechanical properties

Determining how the following variables affect the strength of a cardboard shelf: type of glue, lamination method, shape, support method

Design

Repairing and redesigning bags

Designing cushioning systems for a fragile product

Creating portable storage units for transporting shoebox dioramas

Designing and making cardboard shelves for the classroom

Analyzing and Designing Packaging

Technology in the Early Childhood Classroom *by Theresa Luongo*

Most children are in my class for two full school years. They enter at approximately four years of age in pre-kindergarten, and stay with me until the end of kindergarten, when they are approximately six years old. This school year (1998-99) my class is made up of two-thirds kindergarten students and only one-third pre-kindergarten, due to budget cuts for pre-K. I have 23 students in my class and a full-time paraprofessional.

My classroom is spacious. The room is divided into work areas where small groups of children engage in a wide range of activities. There is a large rug area where we come together to meet and share work and read. The various areas of the room reflect the choices available to the children during work time: Sand, Water, Math, Discover, Blocks, Dramatic Play, Construction, Painting, Cooking, Clay, Writing, Library and Listening. (See Figure 4-1.) Each day, the classroom transforms into a workshop atmosphere where children work together on projects.

Theresa Luongo is a pre-K/K teacher at Central Park East 2, a small alternative school in East Harlem. Packaging became an ongoing activity in her classroom as her children became intrigued with the properties of boxes, bags, and pump dispensers. Many of the activities were initiated by the children, who found new uses for the pump dispensers and discovered how to modify and repair bags. Theresa came away from these activities with a new appreciation for the potential of technology in the early childhood classroom.

4-1: Theresa's map of her classroom

There are certain universals shared among most early childhood teachers. One is that learning is a social process. Another is that learning is active, not passive. If we want our students to be active learners and use language for real purposes, we need to create classroom environments that encourage this. The classroom itself should be viewed as a workshop or studio where experimentation and dialogue take place among the students. Each day a certain amount of time should be devoted to student projects and experimentation. In my school we call it "Work Time." By setting aside a time each day for students to make choices about where they will work, we are helping them pursue their interests and make choices about their learning.

Initially, Theresa was reluctant to integrate technology into this already rich environment. Her overriding goals are in the areas of language arts, and she was not sure how technology would contribute to those objectives. She writes:

One goal of most early childhood teachers is to promote language and literacy development. As a kindergarten teacher, I didn't see how technology could help me achieve that goal and I was skeptical about the role it would play in my classroom. One of my reservations centered around my feelings of inadequacy about technology in general. I was equally concerned about trying to do something new, in addition to everything that was happening in my classroom already.

As I read more and talked with participants who had been part of *Stuff That Works!* the previous year, I started to broaden my understanding of what technology is. I began to see how technology could fit into what happens naturally in the kindergarten classroom and how important it is to expose young children to technology and their part in it. Technology promotes language development naturally. When children come together to repair something or understand the way a mechanical device works, they are forced to use language for real reasons.

It isn't necessary to set up a Technology Corner or a Technology Center for technology activities to take place in the classroom. I want to stress the importance of using technology as an integral part of what already exists in the classroom.

Theresa sought an aspect of technology that would fit neatly into one of the work areas in her classroom. She wanted her students to explore technology in the context of the things they were already doing. In addition, she was hoping that they would pursue these investigations over the long term, revisiting their own questions as new ideas arose. Theresa decided to introduce pump dispensers—the kind found on lotion and cleanser bottles— into the water table in her classroom.

How Do We Get the Smelly Water Out of the Water Table?

This was something I knew I could fit into an area already established in my classroom, the water table. Our water table has a large basket underneath where various objects are kept. Among those objects was one large plastic pump. During a meeting at the rug area I brought over a pump dispenser (filled with water) and a clear plastic cup. I explained that the children going to the water table would have a chance to use these kinds of pumps. I then passed the pump dispenser and cup around to give the students a chance to use it.

Up to three students at a time may go to the water table. In addition to water toys, several pump dispensers were available. Each day I encouraged different students to choose Water as a Work Time activity. The students began experimenting with the pumps. Initially, the children didn't all know that the tube needed to be in the water. They tried to fill the tube with buckets of water, basters, large spoons, etc., only to turn the pump over and have the water spill out. (See Figure 4-2.)

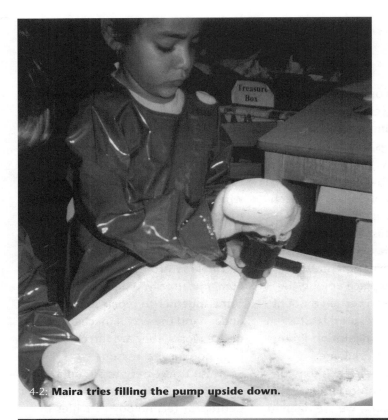

4-2: **Maira tries filling the pump upside down.**

4-3: **The pump collection**

Here is a sample of the talk at the water table:

DIONISIO:
It's bubbling, the water is bubbling. We made it bubble.

JUSTIN (holding a meat baster):
How does this thing work?

THERESA (the teacher):
How do you think it works?

JUSTIN:
I know. You have to push it ?the tip? real hard.

DIONISIO:
You have to squeeze it.

THERESA (holding up the pump):
How does this work?

DIONISIO (turning the pump upside down):
You have to fill it up. Hey, we need some help here. I can't put water in this thing.

JUSTIN:
You should fill it in the sink.

Both students worked on filling the tube upside down, only to have the water spill out when it was turned over. Finally, they decided to submerge the pump's tube in the water.

I asked the students to start looking for pumps at home. The one deal I've made with the class is, "If you bring in a pump, you will definitely get a chance to go to the water table during Work Time that day." Now we have a rather extensive pump collection that keeps growing. (See Figure 4-3.)

Once I felt the students had had enough time to play with and explore the pumps, I began our experiment.

The students selected three pumps of various sizes and labeled them #1, #2, and #3, and also numbered three cups. Then, taking turns, each child operated a pump just once, with the flow directed into the cup with the same number. (See Figure 4-4.) The question was "Which pump will pump the most water?"

Theresa continues:

With each test, we would record the results of our experiment and the date on a chart. (See Figure 4-5.) After several trials, we started testing three different pumps labeled #4, #5, and #6. These pumps are closer in size to one another than the pumps used in the first trials.

4-5: **Pump data chart**

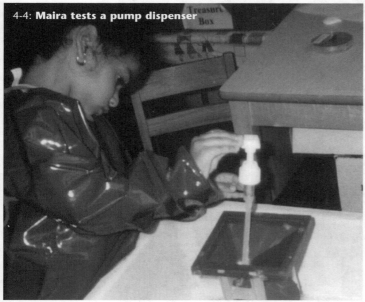

4-4: **Maira tests a pump dispenser**

As we continue to test our pumps, the students are asking and answering their own questions about how the pumps work. Some are convinced the bigger the pump, the more water it will be able to pump. Others aren't sure if that's true.

In the course of these investigations, some children came to see the pumps as useful tools, as well as interesting objects to experiment with. There was a real problem that the pumps could solve: "How do we get the smelly water out of the water table?" They recognized not only that the pumps could do the job, but also that the large pumps could do it faster.

One student actually tried using a spray pump to empty the water table. Now, when it's time to clean the water table, the students are using the large pumps and taking turns doing the pumping. The study of pumps and they way they work is now part of our classroom environment.

Theresa was very pleased with this sequence of events. The study of pumps seemed to fulfill her goal of stimulating language development. As an added and unexpected bonus, her students had discovered for themselves that they could solve real problems using their newly developed knowledge of these devices.

The study of bags proved to be even more fruitful. As with the pump dispensers, Theresa integrated this activity into an existing work area— this time the block area.

From Bag Testing to Bag Repair and Redesign

We began in the block area with an activity designed to test the strength of two small shopping bags. During a group meeting before Work Time, I asked the students who chose Blocks to see how many blocks they could fit into the two bags and to find out which bag could hold more. I asked the students to work as a group to fill each of the bags. (See Figure 4-6.)

4-6: Ronald and Langston pack a bag with blocks.

Once one of the bags broke, I talked about it first with the small group and then with the class. I wanted them to investigate which part of the bag broke. (See Figure 4-7.)

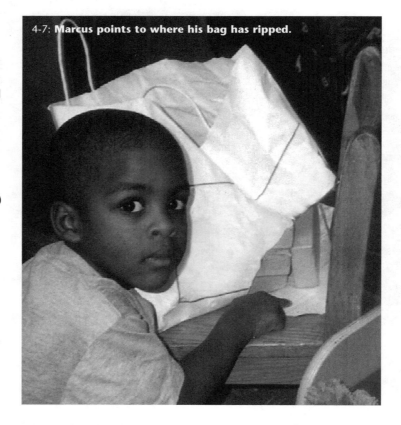
4-7: Marcus points to where his bag has ripped.

This too has taken on its own life in the classroom. It's interesting seeing the rules they make up for packing and unpacking the bag, as the following dialogue illustrates:

RONALD:
We got to do it this way.

LANGSTON:
I know what I'm gonna do...

RONALD:
You gotta share the blocks.

LANGSTON:
I am sharing.

DANIELLE:
First we fill this bag. Who's helping me? Bria, get the blocks down. Bria, I'm getting the blocks down.

LANGSTON:
Look. Hey look! I'm almost finished with my bag.

RONALD:
You can't put them in like that. They won't fit like that.

LANGSTON:
I'm finished. Want to see?

When both bags were filled up, the children tried carrying them around the classroom to show the other students. I asked them to find out how many blocks each bag had. We never really found out. Instead, the objective became just taking the blocks out carefully.

DANIELLE:
We need to make a big pile. Theresa likes it when it's in a pile.

RONALD:
I'll help hold them.

BRIA (counting each block as it's taken out of the bag):
One, two three, four... Put them on top of each other.

RONALD:
We need to take turns. We all need only one block at a time.

LANGSTON:
Yeah, like I'm doing.

Once the bag was empty, they lost count, and proceeded to fill the bag again until the handle broke off one of the bags.

As with the pump explorations, the students took the bag testing in a direction Theresa hadn't anticipated. They decided to repair the broken bags!

After the handle ripped off one of our shopping bags, Langston, a pre-K student, informed me that he knew how to fix it. I suggested that he go to the Construction area and repair it. Langston took out tape and began mending the bag. Mariah (who's also in pre-K) came over to find out what was going on. She stayed and helped Langston fix the shopping bag. (See Figure 4-8.) Soon, another bag ripped in the Blocks area. This one tore on the side seam. Langston and Mariah carefully worked to repair the shopping bag. (See Figure 4-9.)

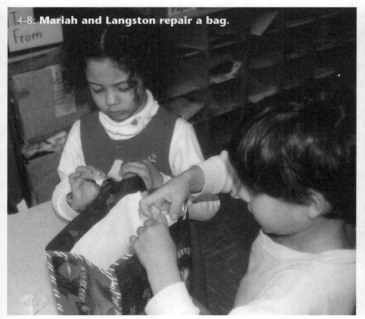

4-8: **Mariah and Langston repair a bag.**

4-9: **Langston shows the repaired bag.**

I asked them how I could turn my small brown paper lunch bag into a shopping bag. Here's the advice they gave me:

MARIAH:
You need string, you know.

LANGSTON:
Yeah!

I took their advice and took out string and scissors. Langston began folding down the edge of the shopping bag, which was something he had observed on the larger shopping bags.

MARIAH:
We need glue.

I suggested putting holes in the bag and asked if they thought that might work. Both Langston and Mariah thought it was a good idea. After I punched the holes, Mariah worked with the string to make knots and handles. (See Figure 4-10.) When they were finished, they brought it over to the Block area.

MARIAH:
We're not putting blocks in this bag. Do you know why?

THERESA:
Why?

MARIAH:
This bag will break.

As a result of what Langston and Mariah did, bag testing and especially bag repair caught on. They became among the most popular Work Time activities in Theresa's class.

4-10: **Mariah creates a shopping bag out of a small brown lunch bag.**

Now the students are figuring out how to repair the bags and discovering where the bags tear most frequently. The children have the opportunity to continue to test the bags over and over again and repair the bags over and over again. Only by trial-and-error will the students discover which tape works best or what will help strengthen the various bags. Each day more children seem to want to go to the Block area and test the bags so they can get a chance to repair them.

Recently a child in the class gave out party bags, which were small shopping bags, to each student in the class. Following the party, students started bringing the bags back to school to put in the Block area.

They also feel empowered by being able to fix something. Bag repair has led to children repairing other things in the room. They must think, "Well, we can fix those paper bags. Why not fix these torn book covers?"

After Taylor, another pre-K student, witnessed Langston and Mariah taping a shopping bag, she informed me that she wants to go to the Construction area so she can repair the torn cover on one of our big books. This will lead to numerous technology opportunities in the classroom. What if our Construction area also became a repair station for broken objects and materials in the classroom? I can't help but feel excited for the child who breaks or tears something, only to learn it can probably be fixed. And even if it can't be repaired, it's worth trying. Think of the message it sends! We live in a throw-away society. Suddenly we find children starting to repair things rather than just asking for new ones.

In her concluding reflections, Theresa discusses the potential for technology in the early childhood classroom:

Technology becomes real for early childhood students and teachers when it's used for real purposes. When children in my class go to the Construction area and figure out how to repair a bag they used in the Block area, they are doing many things:

1. They must work together to figure out what the problem is. Where did the bag tear?
2. They must negotiate how they will fix it.
3. They must test the materials they will use to repair the bag.
4. Most important, it was their choice to repair the bag, so they feel vested in what they are doing.

My goal is to provide a classroom environment that allows children to make choices about the activities they will participate in. Some children wish to experiment with things until they know the materials inside-out and feel they have mastered them. Others like to try new things all the time. The classroom needs to allow for the individual learning style of each student.

My role as the teacher is to find a point of entry for each child, one that supports his or her needs and allows for differences. Now I see technology as an integral part of my classroom and curriculum. The children have given technology it's own life in the classroom and I support it. I began by feeling cautious with a single pump, but now I can't imagine my classroom without technology.

Packaging as Raw Material for Language Arts, Science Process, and Math *by Verona Williams*

Verona Williams is a third-grade teacher in the South Bronx, New York City. Her goals for her unit on packaging were to develop her students' abilities in language arts and quantitative reasoning. She began by having students define "package" and classify bags and packages and explain their uses. The children's own questions and suggestions eventually led to some systematic testing of bags. Verona's story features her assessment of individual children's strengths and weaknesses, based on their writings about packaging. Verona used packaging as a theme for teaching a variety of subjects, including language arts, science, and math. She began by having her students try to explain the word "packaging." Here is her account:

What Is Packaging?

I asked the students to brainstorm in cooperative groups and answer the following:

- List all different types of packaging things you know.
- What is packaging?

The class was divided into seven groups. I gave each group three packages to study. I asked them to define "packaging" and to list everything they thought had anything to do with packaging. They listed things found in packages as well as things that are considered packages. (See Figure 4-11.)

Some of their definitions of packaging were:

I think it is when you pack things and boots, toys.

Packaging is when you put something in a bag.

I think that packaging is when you shop for shoes, hats, lunch.

Packaging is when you put stuff in a plastic bag.

Each group shared their list of different types of packages. Below is a summary of their lists:

- different foods that come in packages (15 instances)
- toys (6)
- gifts (3)
- plastic bags (3)
- bags (2)
- clothes (since they cover the body)
- book bags

4-11: **What is packaging?**

Date 1/12/97

Subject: **Packaging**
Brain storm

1. **What is packaging?**

Packaging is like a suitcase when you go somewhere you put clothes in a package

2. **List all the different types of packaging things you know.**

bay   totor   Christmas Plastic bags
Paper   earring   gift   tissues
gerensle Chains   Cheese   Paper
books   toys   groceries Rose
Clothing hershey   food   pizza
         shoes   bags

3. **Study the items on your desk. Write what you notice about them.**
   a) Object name _bags_ and draw it.

This bag has a handle. The handle is in the middle of the bag

When asked why "food" was given as a package type, several students said that it belongs because food is bought in different wrappings. Even though they listed many items found in cans and boxes, they did not seem to associate these as types of packages.

I then asked the whole class, "From this activity, did we learn what packaging is?" Some of the answers were:

JUAN:
Yes. Packaging is when you put a lot of stuff in a bag, then put it in a carriage.

MARADIS:
Packaging is when you go to the store and you buy things, then put them in a bag.

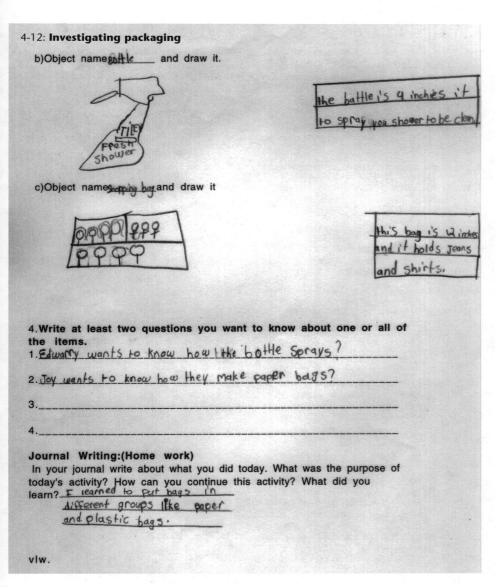

4-12: **Investigating packaging**

   b)Object name bottle ____ and draw it.

the bottle is 4 inches it to spray you shower to be clean,

   c)Object name shopping bag and draw it

this bag is 12 inches and it holds jeans and shirts.

4.Write at least two questions you want to know about one or all of the items.

1. Edward wants to know how l the bottle Sprays? _____

2. Joy wants to know how they make paper bags? _____

3._____

4._____

Journal Writing:(Home work)
 In your journal write about what you did today. What was the purpose of today's activity? How can you continue this activity? What did you learn? I learned to put bags in different groups like paper and plastic bags.

vlw.

Verona asked her students what additional questions about packaging had come up in the groups. For homework, she asked them to make a journal entry about the activity. (See Figure 4-12.) However, Verona felt that some of the children had missed the point of the activity.

I assigned this activity to assess the students' comprehension of what they had done, and to find out whether or not they understood the concept of packaging. One group wrote the following:

We measured things. We put things in the box and bag... We draw it nice and neat.

They were not clear about the purpose of the activity nor did they seem to have a concrete understanding of packaging.

I had also asked the students to bring in packages, bags, boxes, or anything else they thought was a package. This was another way of informally assessing what their concept of packaging was. The students brought in bags only; no one brought in bottles or boxes. Based on this, I decided to narrow the investigations to paper and plastic bags.

MICHAEL:
Packaging is when you tell us to pack up.

JENNIFER:
Packaging is when you pack for yourself and someone else.

After gathering this information, I realized many of the groups still thought packaging was putting things in a bag. I wanted them to understand packaging as a type of container or holder of things. Some understood this and said packaging was a bag for something from a store, especially groceries or clothes. However, they only saw packaging in terms of bags, although I had distributed other types of packages, such as bottles, cartons, and boxes.

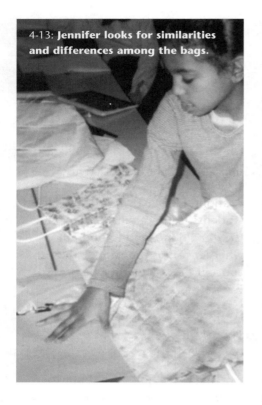

4-13: **Jennifer looks for similarities and differences among the bags.**

Classifying Things

Verona decided to have her students classify bags. She felt that this activity would help them develop the concept of a category, reinforce some math skills, and also help them focus on the aspect of packaging that seemed most familiar to them. First, however, she felt that they needed to learn what it means to classify. To develop this concept, she called to the front of the room a group of students who were all wearing the same colors. The students who were still sitting had to guess what all the students in front had in common. She then repeated this game with some other characteristics, such as students wearing uniforms, female students, and tall students.

Once they were able to see what categories are, Verona went on to the bag activity. She gave each group a variety of bags, and asked them to think about the things about them that make them similar or different. They were to explain the basis for their categories on worksheets, and also construct charts showing how many bags were in each category. For Verona, this activity offered a rich opportunity to assess children's thinking:

4-14: **Germary finds paper and plastic bags as well as big and little bags.**

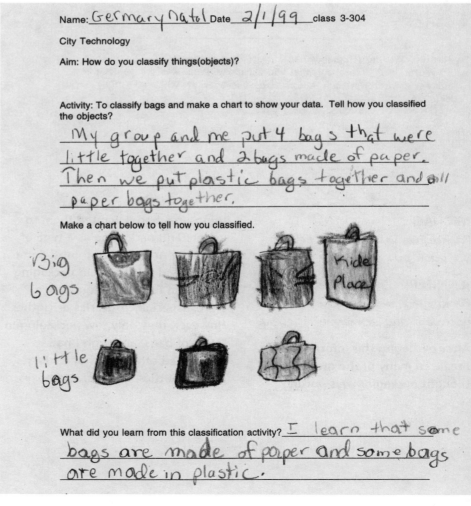

Name: Germary Natal Date 2/1/99 class 3-304

City Technology

Aim: How do you classify things(objects)?

Activity: To classify bags and make a chart to show your data. Tell how you classified the objects?

My group and me put 4 bags that were little together and 2 bags made of paper. Then we put plastic bags together and all paper bags together.

Make a chart below to tell how you classified.

Big bags

little bags

What did you learn from this classification activity? I learn that some bags are made of paper and some bags are made in plastic.

4-15: Maria's bags are dark and big or light and small.

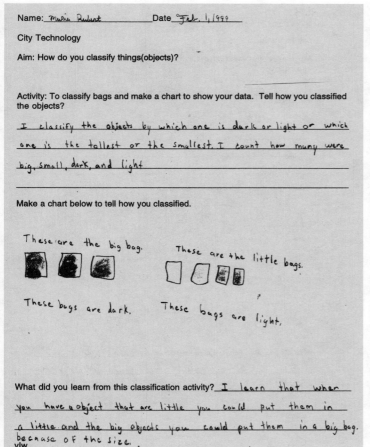

Name: Maria Redunt   Date: Feb. 1, 1999

City Technology

Aim: How do you classify things(objects)?

Activity: To classify bags and make a chart to show your data. Tell how you classified the objects?

I classify the objects by which one is dark or light or which one is the tallest or the smallest. I count how many were big, small, dark, and light

Make a chart below to tell how you classified.

These are the big bag.   These are the little bags.

These bags are dark.   These bags are light.

What did you learn from this classification activity? I learn that when you have a object that are little you could put them in a little and the big objects you could put them in a big bag. because of the size.
vlw

4-16: Fun with bags

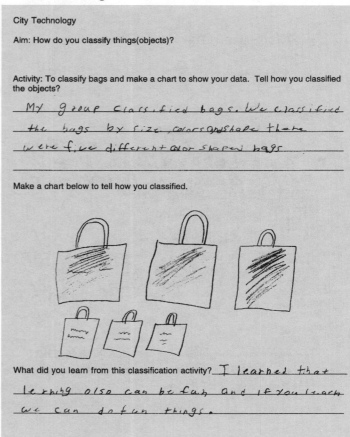

City Technology

Aim: How do you classify things(objects)?

Activity: To classify bags and make a chart to show your data. Tell how you classified the objects?

My group classified bags. We classified the bags by size, colors and shape there were five different color shaped bags

Make a chart below to tell how you classified.

What did you learn from this classification activity? I learned that lerhig olso can be fun and if you teach we can do fun things.

Each student had his or her own handout on which to record the data. Several students simply drew their bags, rather than create a chart, as I had requested. Germary struggles with writing and she has a hard time trying to express herself. She first describes two unrelated classifications: "little" and "paper." Then she says her group had grouped the bags according to whether they were plastic or paper. Her chart shows "big bags" and "little bags." Then at the end of her statement she again referrs to paper and plastic. She was not clear as to how the size categories and the paper-or-plastic categories are related. (See Figure 4-14).

Maria listed the categories of dark or light and big or small, but did not distinguish between these two ways of categorizing. (See Figure 4-15.) One student classified her bags by size, color, and shape. However, the only information she could convey in her chart was the relative sizes of the bags. (See Figure 4-16.)

Michael gave detailed descriptions of the different bags he had, but neglected to sort them into cate-gories. He discussed how different bags are used for different purposes, but did not explain how he had learned these facts. (See Figure 4-17.)

Siara classified by color but gave no details about the process she used. The data on her graph was incorrect. I spoke with her about the numbers and she realized that her group had never had that many bags. At least she was aware that a graph would be an easy way to represent one type of variable, such as color. (See Figure 4-18.)

4-17: Michael decides what bags are used for.

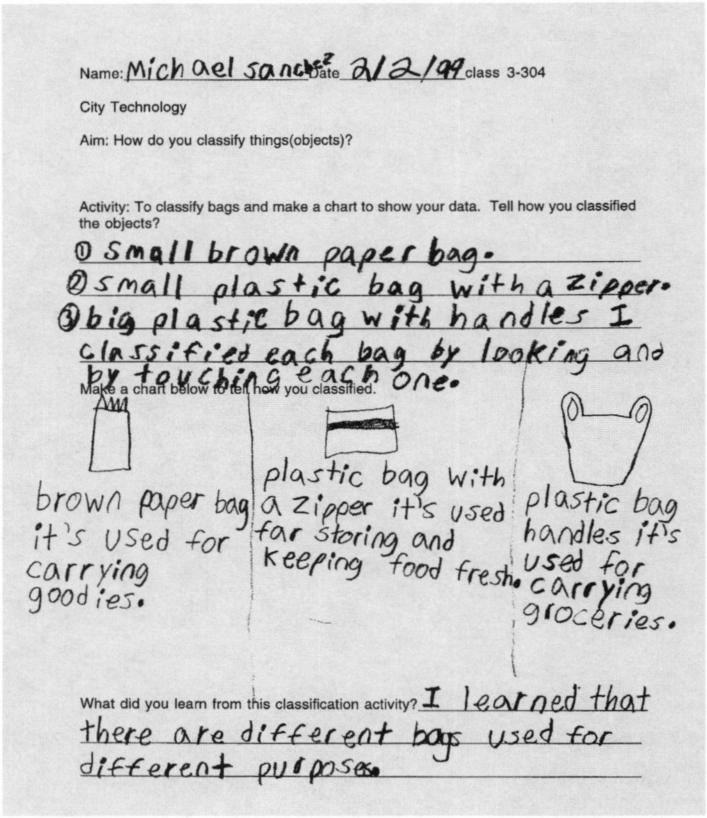

Name: Michael sance Date 2/2/99 class 3-304

City Technology

Aim: How do you classify things(objects)?

Activity: To classify bags and make a chart to show your data. Tell how you classified the objects?

① small brown paper bag.
② small plastic bag with a zipper.
③ big plastic bag with handles I classified each bag by looking and by touching each one.

Make a chart below to tell how you classified.

brown paper bag it's used for carrying goodies.

plastic bag with a zipper it's used far storing and keeping food fresh.

plastic bag handles it's used for carrying groceries.

What did you learn from this classification activity? I learned that there are different bags used for different purposes.

4-18: Siara graphs according to color.

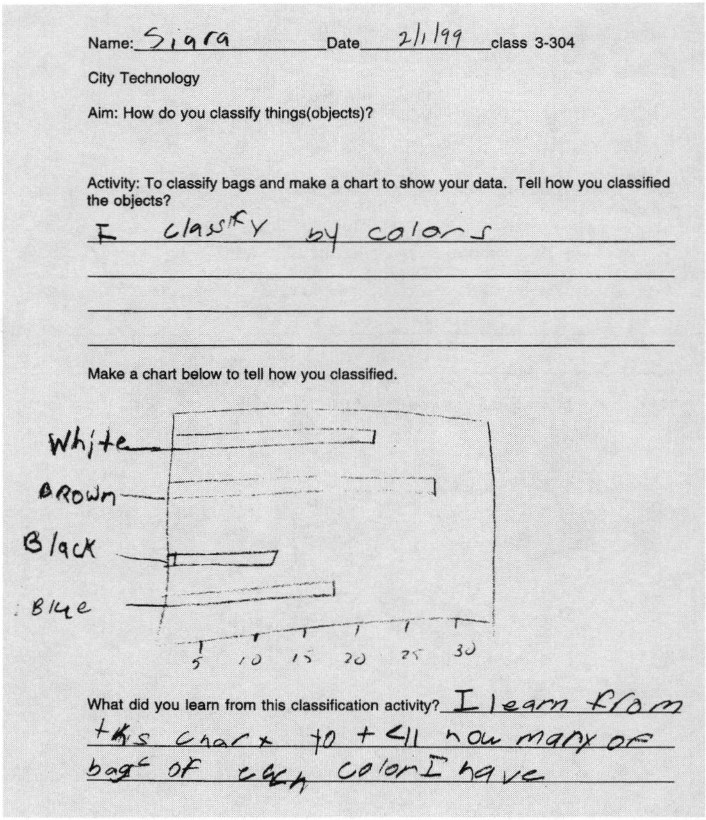

Name: Siara Date 2/1/99 class 3-304

City Technology

Aim: How do you classify things(objects)?

Activity: To classify bags and make a chart to show your data. Tell how you classified the objects?

I classify by colors

Make a chart below to tell how you classified.

White
Brown
Black
Blue

5 10 15 20 25 30

What did you learn from this classification activity? I learn from this chart to tell how many of bags of each color I have

4-19: Kamina classifies dolls instead of bags

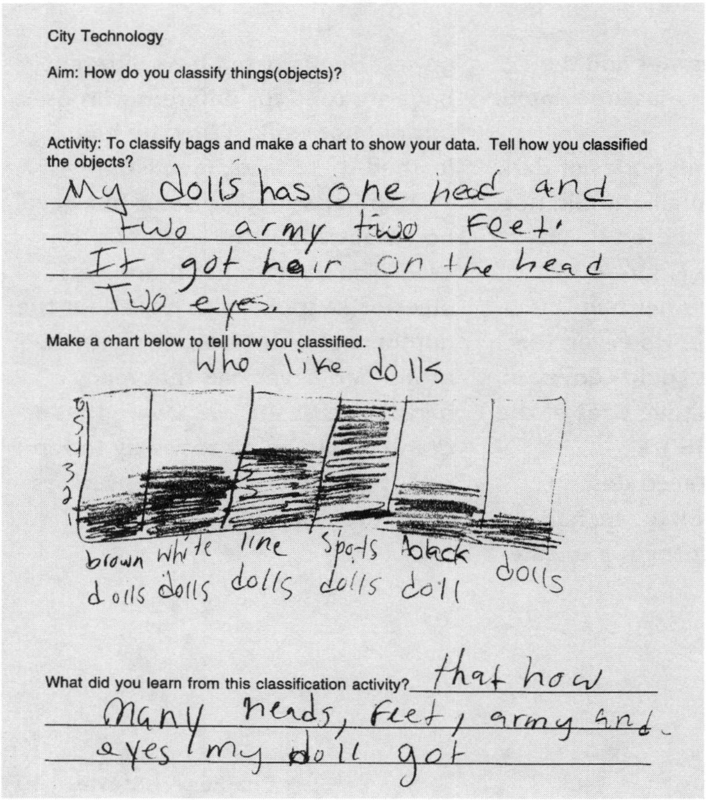

City Technology

Aim: How do you classify things(objects)?

Activity: To classify bags and make a chart to show your data. Tell how you classified the objects?

My dolls has one head and two army two feet. It got hair on the head Two eyes.

Make a chart below to tell how you classified.

Who like dolls

brown dolls white dolls line dolls spots dolls black doll dolls

What did you learn from this classification activity? that how many heads, feet, army and eyes my doll got

Kamina's data had nothing to do with the bag activity. When I asked her where she had gotten her information, she said she had decided to graph the dolls she has at home. (See Figure 4-19.)

In her reflections on this activity, Verona felt that the students had been successful in grasping the concept of categorization, but were not clear about how to represent their data. She felt that this activity could serve as an excellent lead-in to making charts and graphs of data:

The students learned the various ways to classify— by color, size, shape, and material. Some of them tried to represent their findings by drawing, charting, or graphing. However, these were first-time efforts. Combining this activity with math lessons about charts, graphs, and quantifying information would provide a frame of reference.

In reflecting on the design of the activity, Verona felt that the handouts should have been completed by groups rather than by individual students. Individual classifying activities could have followed once students had worked on classifying within groups:

The students realized that discussing their findings as a group, and using one data sheet per group, would have been more productive. They saw how well teamwork had worked in the lesson on defining "packaging." Immediate team feedback and consensus had occurred when each group had only one handout for recording and representing their data. An individual classifying activity would work better afterwards, allowing students to pursue their own interests.

Verona was intrigued by the variety of categories the students used. In retrospect, she felt that she should have focused on the different ways of classifying. Although Verona saw room for improvement in what she had done, it was clear that the study of bags had engaged her students' imaginations.

Several weeks later, the class revisited their investigation of bags. Verona asked her students to study the physical properties of their bags, and come up with some questions they would like to explore further. Maria came up not only with a question, but also with some hypotheses.

Maria was the only student who had written in her journal. She began by asking the question, "How do you test a bag?" She then answered it: "I will test my bag by putting a lot of hard rocks to see if it breaks easily. Also I could say that the bag that has a plastic handle could last longer than the one that has the paper handle."

Maria's question became one for the whole class. They decided that they would test the strengths of their bags. Because no rocks were available, they decided to use books. A question arose about whether the books for testing one bag had to be the same as those used for the others. Verona put this question to the class.

I asked, "Why would you use one type of weight to test the strength of the bags?" Maria said, "... so we could use the same books with all the bags to see which one was the strongest bag. If two bags were the same size and had handles, we should keep the weights the same to really see if one is stronger than the other."

Without using words like "fair test" and "control of variables," Maria had grasped these concepts in the context of a question that was meaningful to her. Her intellectual engagement in this activity continued as they actually tested the bags. In reporting the test results, Maria offered her own explanations of what had happened:

Paper bag: The bag had a handle. We will use big red books. I think we will put 6 books in the paper bag to see if it break. The 6 books that we put broke the bag on the front. The paper bag did not hold 6 big books because the bag was made out of paper and paper is easy to break.

Plastic bag: If you put 5 or 6 or 4 books it will break because when you carry it and there are a lot of books the handles become slippery from your hands because it is too much book. The plastic bag could hold 2, 1 or 3 books. The heavy book strains the bag.

Although Verona felt that this unit had stretched her students' imaginations, she also saw that they needed more help in recording and organizing their data. She added that there should have been more analysis of how the bags had failed.

A closer look at where and why the bags ripped may assist the students in looking at the function of particular bags. Are they doing the job they were designed to do? Are there other ways to test their function? An extended activity would then be to design a more appropriate bag.

Investigating Fruit and Cushioning
by Roslyn Odinga

Roslyn decided to implement extended packaging units with two of her classes. Her second-graders looked at natural packaging, and her fourth-graders explored cushioning.

Natural Packaging

Roslyn began with a simple idea: explore the "natural packages" that surround most kinds of fruit. When she started this project, she had little idea of the other connections her second-graders would make.

April 14
I asked the children what packages are and when something is a package. I introduced them to different types of packaging, e.g., packaged cereal, rice, ketchup, beverages, and gift-wrapped boxes. (See Figure 4-20.) I asked: "What happens when you go to the produce area of the supermarket to purchase fruit? Is it packaged or contained in something?"

Roslyn Odinga is a science cluster teacher at Community Elementary School 126, a large school in the South Bronx, New York City. Her schedule requires her to meet with 25 different classes during the week for 45 minutes each. The classes range from second to fifth grade. Sometimes, due to tests or special programs, some of these meetings are missed, and she does not see the same group for two or three weeks in a row. Under these circumstances, Roslyn finds it very difficult to maintain the continuity of a long-term project, or even to complete a meaningful activity in a 45-minute time slot.

4-20: **Various types of packaging**

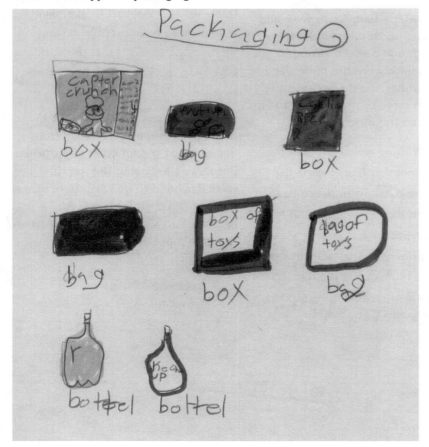

I introduced the idea that some things we purchase are naturally packaged—i.e., the packages are not made by humans. I showed the children examples of fresh fruit. They examined oranges, apples, grapes, bananas, mangoes, kiwi fruits, a cantaloupe, and a coconut. I had them pay special attention to the covering of each of these fruits. The students were very excited about actually working with fruits. I got a lot of positive responses as to what natural packaging is. We talked about our bodies and how they are actually packaged by the skin.

April 21

I decided I had to take them on a field trip to the supermarket. We went to the produce section and saw all the different types of fruits and realized that fruits don't necessarily come in bags. Jiovanni and Jennifer M. were unaware of certain types of fruit. Kiwi and cantaloupe were complete mysteries to them.

I also noticed that the children had no sense of prices. They were amazed that apples were 99 cents a pound. Then we got into a discussion of what a pound is. We used the scale to determine how many pounds of each fruit we had. Another question was, "If they sell five oranges for $1.00, how much would each one be?" The students really enjoyed their trip.

It is remarkable how much we take for granted about children! One of the striking features of "real-world" activities, such as going to the supermarket and looking at fruit, is how they expose aspects of children's understanding that are rarely touched by traditional instruction. Roslyn's next activity was similar, in that it required her students to come up with their own vocabulary for describing textures.

April 28

Students examined fruit samples, paying special attention to the textures of the skins. I got answers like "smooth," "rough," "bumpy" and "hairy." (See Figure 4-21.)

We discussed how packaging protects fruits. I cut the fruits so that the students could observe the differences between the outsides and the insides. They were noticing colors, and also seeing that under the skin is the actual fruit. We then went further, and classified by whether the skins are edible or inedible.

There also turned out to be a big discussion about coconuts. How do coconut trees grow, since coconuts don't have any seeds? We found out that when a young coconut falls from the tree, it's still green, and is planted. So, my students came to the conclusion that a coconut is actually a seed. I could not dispute them! I was really amazed by their intellect.

4-21: **Classifying fruit by texture**

HOW DO I FEEL?

Directions: Write the names of different fruits in each box. Put a check in the correct column if the covering of the fruit is Smooth or Rough.

Name of Fruit	Smooth	Rough	Other
mangoe	✓		
strawber.		✓	
Banana	✓		
Kiwi		✓	hairy
Apple	✓		
pear	✓		

May 18
Today is Fruit Salad Day. It's like a party. We talk while I make the fruit salad. Some children taste fruits they've never tasted before. "No coconut!" they shouted in unison. They didn't want to eat that giant seed, plus they were afraid it would get in their teeth. I was also questioned as to why a banana will get brown so fast. This is going to lead us into another type of inquiry...

Which Type of Cushioning Is Best for an Egg?

Roslyn conducted a different kind of packaging unit with a fourth-grade group. Because they were older, she wanted them to do a controlled experiment and collect and present quantitative data. This activity also involved some design, because the cushioning materials and their arrangement were not specified. As with "Natural Packaging," the students took this activity much further than Roslyn had anticipated.

Day 1
I showed the students a dozen eggs and demonstrated how they were packaged in a carton with an individual compartment for each egg. Pretending to do it by accident, I dropped the carton of eggs on the floor and of course several broke. I once again showed the students the carton and asked them how the eggs could be packaged more effectively. I then presented students with the challenge:

"Package an egg so that it won't break when dropped from different heights."

I let them preview the materials that we were going to work with. These included foam rubber, Styrofoam, newspaper, bubble envelopes, egg cartons, Styrofoam peanuts, Enviro-bubble (a brand of bubble wrap), cloth, towels, masking tape, packaging tape, paper, pencils, markers, journals, and small milk containers.

Of course I expected some feedback, because I want them to have a VOICE. They were upset about having to work with the small milk containers, which they said were too small. Theodore pointed out that when you buy a carton of eggs at the supermarket, of course the packaging is larger. They just section off spaces for the eggs! I couldn't dispute that fact, so we changed the size of the packaging. I allowed them to use shoeboxes, which were 8" x 10", and whatever else they could bring in from home.

Roslyn recognized the need to open up the materials list to include anything the students could find. In this way, she encouraged them to be resourceful in solving a technology problem: What would make good cushioning material? She also gave them a greater stake in the project by turning a piece of the problem over to them.

The students also looked closely at conventional cushioning supplies, such as Styrofoam peanuts and bubble wrap, which they referred to by its brand name, "Enviro-bubble." They became involved in figuring out what these materials are made of and how they are made. Focusing on these common artifacts strongly stimulated their curiosity.

Day 2
I asked them to write down their observations about different kinds of cushioning material. After students examined all the materials, there was a discussion to determine a name for each item. Donald took it upon himself to name the two types of foam (foam rubber and Styrofoam): "soft" and "hard." With the Styrofoam peanuts, he had a little problem. He said that although they looked like foam, they had a different type of feel to them. (See Figure 4-22.)

"Enviro-bubble"—that's the phrase of the day. "Enviro-bubble this! Enviro-bubble that!" There were a lot of questions as to how they get the air inside of the bubbles. Is it processed in a certain type of machine that has allotted spaces for air? Or, is it blown up and then pressed down, which allows the air to stay in specified places? I LOVE THESE TYPES OF QUESTIONS. Now they have research to do!

Day 3
Although the students had worked in groups to do their initial observations, they decided to work individually on the actual packages. The students first tried packaging their eggs individually using shoeboxes and other types of material such as aluminum foil, Pampers, swatches of clothes, cotton balls, Styrofoam, etc.

They then tested their packages by having them dropped from a designated height. We then opened the boxes and talked about the types of packaging that were used.

4-22: Donald describes the cushioning materials

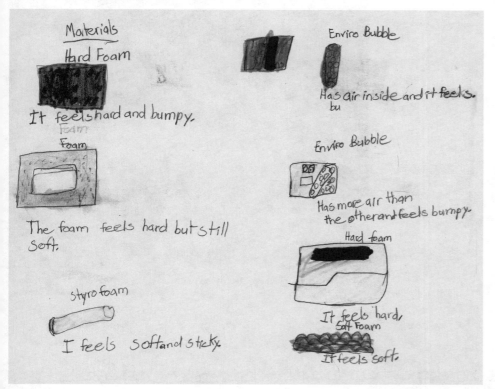

Materials
Hard Foam

It feels hard and bumpy.

Foam

The foam feels hard but still soft.

Styrofoam

I feels soft and sticky.

Enviro Bubble

Has air inside and it feels.
bu

Enviro Bubble

Has more air than the other and feels bumpy.

Hard foam

It feels hard,
Soft Foam

It feels soft.

4-23: **Theodore with his egg-cushioning box**

Theodore's egg broke, because he failed to tape down the top of his box. (See Figure 4-23.) Grace's egg was wrapped in aluminum foil, with an underlayer consisting of a wad of paper towels. It also broke. The other drops were all successful.

Day 4
After the preliminary work with shoeboxes, I could see they were now ready to go on to the formal challenge. Each child was to package the eggs in three small milk cartons, in any way he or she wanted, and compare the results:

- Carton #1 was used for Styrofoam peanuts;
- Carton #2 had newspaper; and
- Carton #3 contained Enviro-bubble.

Basically all of the cartons with peanuts were packaged the same way: peanuts on the bottom, egg in the middle, and peanuts on top, with one or two peanuts on the sides. Carton #2 (newspaper) was a little more interesting. One had newspaper on the bottom, egg in the middle, and newspaper on the sides and top. Patrice wrapped her egg first in about one-and-a-half sheets of newspaper. Then she cut newspaper squares and put some in the bottom of the carton. She put her egg in the middle and then squished a little newspaper on top. For Carton #3 (Enviro-bubble), Patrice used the same process as she had used for newspaper. Theodore once again neglected to tape his top, so his packaging wasn't successful. Donald layered his: Enviro-bubble on the bottom, egg in the middle, Enviro-bubble on the top and sides. We were then ready for the actual testing.

Day 5

We started testing by dropping the eggs from 50 cm., first testing Carton #1 to see if there would be any damage. (See Figure 4-24.). At each try we increased the height of the drop by 25 cm. The largest drop was 150 cm. We followed the same process with Cartons #2 and #3. (See Figure 4-25.) The students found the Enviro-bubble to be the best packaging material for the egg. They seemed to feel that the air pockets provided greater protection.

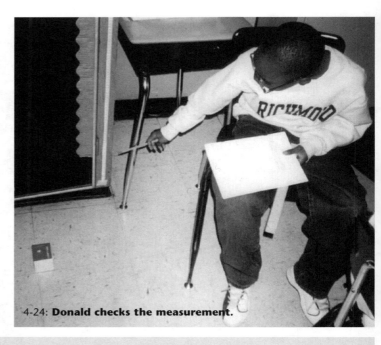

4-24: **Donald checks the measurement.**

4-25: **Data from the egg-drop tests**

Name	Height	Type of material	Condition of Egg
Pat	150cm.	peanuts	○
Donald	150cm.	peanuts	⊜
Theodore	150cm	peanuts	⊜
Angeridno	75cm	peanuts	⊜
Pat	150cm	Bubble	○
Donald	75cm	Bubble	⊜
Theordore	150cm	Bubble	⊜
Angeridno	150cm	Bubble	⊜
Pat	150cm	Newspaper	○
Donald	150cm	Newspaper	⊜
Theordore	75cm	Newspaper	○
Angeridno	0cm	Newspaper	⊘ Couldn't find

Roslyn was very pleased with this unit, which had engaged her students in both a design activity and a controlled experiment. The chart in Figure 4-25 presents the data in a novel way. However, Roslyn felt that it was unreasonable to try to cut up this type of project into 45-minute periods. She resolved to do something about her schedule.

I decided to speak to the principal about extending the science period for next year. I figured that if the classes had a double period with me, and went in clusters (e.g., rotating through the science room for eight-week stretches), they would learn more. As an alternative, I suggested that I could teach one period alone, allowing the regular teacher his or her prep, and then we could co-teach during the next period.

The only response I got was that either plan would be impossible. Our principal will be leaving in June; maybe the new principal will be amenable to the idea. I will try again.

Fifth-Graders Test Bags
by Minerva Rivera

Many of Minerva's concerns centered around the social development of her students, and their ability to work in groups. At the same time, she wanted them to expand their curiosity about the world around them, and develop ways of answering their own questions. It was in this spirit that she conducted her unit on bags.

Minerva Rivera taught the only fifth-grade class at Harbor Academy, a small alternative school in East Harlem, New York City. Minerva was a second-year teacher, without much experience in teaching inquiry science. She wanted her students to develop their own systematic investigations of how different types of bags compare. In her reflections, she discusses how these activities contributed to her own thinking about science teaching.

Day 1
The students worked in groups of four for approximately 45 minutes. The following materials were provided to each group: lab paper, pencil, plastic and paper bags, rulers, tape measures, construction paper, and markers.

I asked each group to examine their bags for differences as well as similarities in both construction and function. I encouraged them to write down any questions that might come up during the course of their investigation. They were also required to think of some ways to test a few of the qualities they listed on their lab sheets.

Much time was spent in getting the students to discuss what they saw and to describe it in written form. Many students felt that writing their observations was the hardest part. I sat down next to a few students and helped them focus by asking a few questions:

MS. R:
What are some of the similarities that you've noted so far on your paper?

BIANCA:
Well, they all have handles (meaning the plastic bags). And they all made of plastic.

MS. R:
What else do you notice about them?

LATAURA:
The bottom of the bags is the same.

MS. R:
What do you see?

LATAURA:
All the bottoms are sewed the same way. They look like they are sewed straight across.

JONATHAN:
But the handles are different. Some of them are on the sides this way (showing the length of the bag) and the others are this way (showing the width of the bag).

MS. R:
Why do you think that they are made that way?

JANNE:
Maybe it helps the bag be stronger.

MS. R:
What can you do to test this? See if you can find things in this room that will help you do this.

I left that group and went around the room and did the same thing with the other groups. I asked a few questions to get some groups started while other groups were way ahead and had started devising ways of testing the bags. They tried to load as many books as they could into the bags to see if the bags could hold that much weight.

A few of the students did not want to work on the project at all and it took my constant prodding to get them to continue. Since it was difficult to get the whole class motivated to do this project, I decided that it would be better to have three groups working on the packaging project as opposed to the six that I had now.

The other thing I noticed is that the majority of the students had a VERY hard time with the lab sheet I devised. I wound up having to explain the sheet three times to the class and several more times to individual students. This sheet definitely needs to be modified. The students need to get the vocabulary (i.e., "procedure," "materials," etc.), but I need to put the information into simpler terms. Instead of just using the word "Procedure" I could use this instead: "Procedures: the steps you followed in doing the investigation..."; instead of "Materials," "Materials: the items you used in the investigation..." and so on.

Minerva recognized some of the problems with this lesson and corrected them the next day. She was able to divide her class in two. Those children who were interested in the project continued to work with her on packaging, while the other students worked with another teacher. She also provided clearer instructions about how they were to document their work. At the same time, she still left most of the thinking and planning to her students.

She was able to make some of these changes because of the flexibility in scheduling and use of space afforded by a small school. Minerva could allow her students to work in the hallway and use the bathroom as a resource for their investigation.

Day 2
I started out by letting the class know that they were to be split into two groups. I put those who were somewhat enthusiastic the last time into the group that would be working with bags again. Those who had not really been interested stayed in the classroom with Ms. K while I worked out in the hall with the packaging group.

I put up a very simple paper that requested that they look at the similarities and differences among the bags. I asked that they also come up with a way to test the bags that would be slightly different from what they had used the last time. I did not specify what aspect of the bags I wanted them to test because I wanted to leave that to the children. I felt that if I gave them enough time, they would be able to find a variety of things to evaluate.

Sure enough, the first group came up with the idea of testing the strength of paper shopping bags after they get wet. It was so enjoyable to watch these students get all excited about working on this project. These particular students ran to the bathroom with the paper bags. Once they got there, I asked them what they were going to use to measure the amount of water they were to add to the bags.

MS. R:
What kind of measuring tools are you going to use in order to collect your data?

JANNE:
We could use one of those measuring cups in the classroom!

JONATHAN:
Yeah and we could use the baseball to see if the water messes up the bag!

MS. R:
What do you mean by that?

JONATHAN:
I mean that we could use the baseball you have in the classroom to put it inside of the bag to see if it'll make them break.

ALISHA:
Well, I could count to see if the bag lasts long.

JONATHAN:
Okay, then I could pour the water and Janne can hold the bag.

These students found that the bags were not all the same when it came to withstanding the water test. The group spent about thirty minutes gathering their data. They poured a cup of water into the bags and then saw how long it would take for the bag to start leaking. (See Figure 4-26.) They placed the ball in and waited to see if the bag could hold the weight. All of the bags broke easily except for a gift bag that had a cardboard-like piece on the bottom.

ALISHA:

Ms. Rivera, I think that this bag is better than the rest because of this cardboard. It is the thing that helps it hold the ball longer. That's why it took so long to break.

The students were able to work for about an hour and ten minutes without losing momentum. In fact, the groups were so into the work that Ms. K had to remind me that it was time for lunch.

The data from these water tests are shown in Figure 4-27. Another group tested bags using books, in much the same way that Verona's students did. (See page 105.) The third group was hampered by one student who tried to dominate. Minerva felt that the activity had gone well in general, but that she could have provided more direction.

I've learned several things doing this unit. I regret that I did not brainstorm more often. I have to find a better way of documenting the words of my students so that I can have more quotes from them. I am contemplating videotaping them, so that we could analyze the film as a class and have a chance to see what actually happened, as opposed to getting snips and pieces of whatever manages to be recorded. The children truly enjoyed this particular activity and have asked to do it again. I think that next time I would break the class into smaller groups so that the children would have more time to process and investigate without the added pressures of competing with one another.

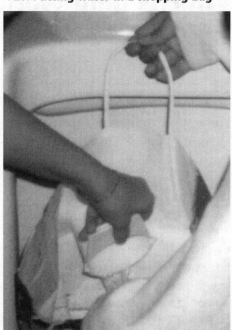

4-26: **Putting water in a shopping bag**

4-27: **Bag water-test data**

	only water	water with a Ball
Bag- 4	Started to dripp at 16sec.	1second started to leak, never broke.
Bag- 3	Started to dripp at 3.5 sec	3 seconds, the bag broke
Bag- 2	Started to dripp at 5 seconds.	60 seconds and 33 seconds, it broke
Bag -1	Started to dripp as soon as the water was poured in	1 minute never brocke.

Which Pump Works Best?
by Christine Smith

Christine made a collection of pump and spray dispensers from the tops of cleaning fluid bottles, soap containers, lotion bottles, etc. She began the unit by distributing a spray pump and a lotion pump to each group and asking them to make and record their observations.

Christine Smith was a science teacher in her second year of teaching, at I.S. 164, a middle school in Washington Heights, New York City. Christine came to the job with a very strong science background, but had serious problems with classroom management. She decided to implement a project on systematic testing of pump and spray dispensers. She felt this unit would both be fun for her sixth graders, and also give them experience in designing and carrying out controlled experiments. She met with this group three times a week, for one double period and two single periods.

October 22 (double period)
I placed the students in groups of 4 or 5. They began the activity by observing the pumps they would later be testing. I gave each group two pumps in a plastic bag, a ruler, paper, and colored pencils. I asked them to observe the pumps and record their data through written notes and drawings. They were to use their five senses to gather data on such topics as size, color, material, and moving parts. Water was also available for them to test with.

Maritza's drawings of her pumps are shown in Figure 4-28. She wrote the following:

The pump is white. The tube looks like if it's foggy inside. The shape is weird. The top looks like a duck's beak and then the bottom is long and thin. The pump sounds like it is very hard inside and you have to push it real hard. Also it sounds like if it got something inside of it that's stuck. The material is made of plastic. That is my choice because when it falls on the floor it doesn't break.

They are the same because they both throw water out. They are both made out of plastic, both of them have tubes that are clear, and you can see the water through.

They are different because one of them is bigger than the other, one throws water straight and the other throws it crooked, and one of them got yellow and white and the other is all white.

4-28: **Maritza's drawings of her pumps**

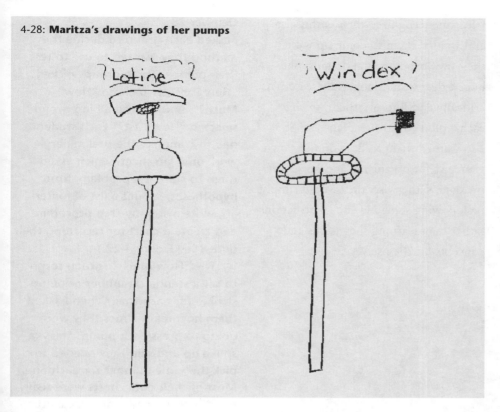

October 26
We discussed the results of the previous day's observations. I asked them to focus on several questions:

- How are the spray pumps different from the "pump" pumps?
- How does each pump work?
- When do we use each kind of pump?

These discussions went really well. The students, for the most part, made excellent observations of the pumps and generated some good thoughts on how the pumps worked:

- Ernie noted that the pumps you push down on are good for thick liquids that you don't want to spray anywhere.
- Jennifer made the observation that the spray pump has adjustments for the spray, but the lotion pump cannot be changed in any way.

The students were intrigued by these common devices, which they had never examined carefully before. They became aware of the resistance that makes the top hard to push down. They also saw the difference in viscosity between the liquids squirted by pump dispensers and those dispersed by spray dispensers. Christine felt that these preliminary observations and discussions served their purpose of preparing the students for designing product tests.

October 26
The focus of our discussion was: "What makes a good pump?" If you are given different liquids like milk, ketchup, laundry soap, water, and oil, how could you test the pumps to see which pump works best for which liquid? I wanted them to come up with different ways to measure the effectiveness of the pumps.

I was impressed with how quickly they were able to come up with different methods of testing the pumps, and with the validity of their ideas. The first idea came from Shirley, who thought the best way to measure a pump was by "how much the liquid came out." When I asked her how she would measure that, she said to put the pump in the liquid and measure how far away the liquid went when pumped or sprayed. The class decided to call this the "How Far?" Method.

We had just done an exercise on reading graduated cylinders, so it was no surprise when Suleyka came up with the idea of using a graduated cylinder to measure how much came out when the pump was pumped. This method got the name "How Much?"

After some more brainstorming, Kathy came up with the idea of measuring how fast the liquid came out. She suggested timing to see how long it took to get the liquid out of the pump.

I asked for any other suggestions, and Giancarlo suggested doing five strokes of the pump, and seeing if any came out. I let him test his idea, but the liquid just went up and down the tube, so then he revised his method: count how many strokes are needed until any liquid comes out. We called this method "How Fast?"

In the groups, the students were sharing their ideas with their peers as they tried to come up with experiment designs. This worked really well, and they produced high quality work. I think that giving each student their own sheet helped in the sharing process because they were all being held accountable for the work. How to make the tests "fair for each pump" was a frequent question and stumbling block.

The students had a surprising understanding of the pumps, and came up with these ideas relatively quickly with little help. Their methods were still missing a few key ideas that were needed to make them valid. However, I decided not to ask them about validity until they had written the procedures down.

Obviously, the students were very engaged in this activity and had a lot of their own questions about the pumps.

Christine recognized their enthusiasm, and wanted them to come up with their own experimental designs. She hoped they would look at these designs critically and improve them, so that each would constitute a fair test. She also wanted them to develop their own strategies for organizing and presenting the data. Otherwise, they would simply be following someone else's instructions rather than learning these skills and concepts for themselves.

October 28
I asked each group to decide the method they wanted to use to test their pumps. Three groups picked "How Far?", two chose "How Much?", and the remaining group selected "How Fast?" Each student had to complete the preliminary work on a lab sheet, which asked them to state the problem, form hypotheses, submit a list of materials, write a step-by-step procedure and create a chart for recording their data. (See Figure 4-29.)

The "How Much?" group forgot to set a standard number of pump strokes for each trial. When I asked them how many times they were going to press each pump, Suleyka spoke up and said they needed to pick the same number for each one. Most of their data charts were really good. For the few that were a little confusing, I asked the students to put in a fake set of numbers. They were then able to come up with a better way of organizing data, once they had a "pretend" data set to work with.

October 29 (double period)
This was the initial test day. It was productive, but extremely chaotic. I underestimated the behavior problems in my class. The main problem was that I thought the other groups would be interested in watching the group that was testing, but I was wrong! In spite of this, the testing went really well. It was extremely messy, but that was to be expected, given the fact that we were spraying stuff around.

4-29: **Testing the pumps**

Pump Lab Sheet

State the Problem:
(What is it you are trying to find out)
how far each liquid goes

Hypothesis:
Which pump do you predict will be best for soap? Pump 1

for milk? Pump 1

for oil? Pump 1

for ketchup? Pump 2

for water? Pump 1

Materials:
What materials will you need?
Ruler, milk, oil, ketchup, water, two pumps
News paper, soap

Procedure:
How will you test the pumps? What steps will you follow for your test?
Put a straight line of numbers with the newspaper by centimeter test all the liquids with one pump at a time. then do the same for the second pump. write all the information and put it in a chart.

Observations:
Create a chart or table to record the data you collect during your experiment. (Please use the back if you need more space.)

how far each liquid goes?		
	Pump 1	Pump 2
Soap	200 cm	40 cm
Oil	205 cm	38 cm
Ketchup	175 cm	21 cm
Water	91 cm	25 cm
Milk	110 cm	40 cm

I also underestimated how far those pumps could disperse liquids. Kids were fooling around a little with the pumps, but for the most part, they LOVED testing. By the end, however, the other groups became bored and rowdy waiting for their turns.

Rumaldo came up to me at the end of class and told me that the effectiveness of a pump depends on the width of the tube. He said that ketchup and laundry soap need a fat tube, but the width doesn't matter for water or milk.

Another interesting discovery had to do with ketchup, which is really hard to spray. The kids discovered that if you use the same pump to spray water first, and then put it right into the ketchup, it will work! When I asked them why this happens, one student thought the water makes the tube more slippery and the others thought the tube just needed to be cleaned out.

I did not see the "How Fast?" group's work, but their results were reasonable, and they had no difficulty doing the test. We ran out of time before the "How Much?" groups could test.

Christine was concerned about the classroom management issues. She hadn't provided enough to do for the groups that weren't testing at the moment, and the two "How Much?" groups hadn't tested yet. Also, she wanted the students who had already tested to summarize what they had done and draw conclusions. She made up a reflective assignment for them to do while the "How Much?" groups tested.

November 2
I came up with a "Conclusion" assignment for the groups who were done with the tests. (See Figure 4-30.) This also allowed me to work with the last two groups. I seated both "How Much?" groups together in a big rectangle with the test materials in the middle. I let them take turns testing while everyone recorded the common results. This worked much better and it was 100% calmer than it had been the day before.

November 5 (double period)
Next, I wanted the students to graph their data. I was surprised to learn that they had just finished graphing in their math class, so they already knew how to use a bar graph to organize their data. We went over a few different ways to set up the axes, and how to define the categories for the graph, but this was really the simplest part of the project. They broke into groups and created their graphs on chart paper. They were excited to have big paper and scented markers. I had a few students who did not work well in groups, so eventually I just gave them paper and let them create their own graphs.

4-30: Some conclusions from the "How Far?" tests

Conclusion
Which pump was best for water? Draw a picture of the pump.

Pump 1

Why do you think so? What about that pump made it better for pumping water?

how the liquid go+ inside the plastic part and came out very far it came the distance of 91 centimeters and that the pump doesn't throwout big drops it throws out small amounts and it goes out very far.

What pump was best for pumping ketchup? Draw a picture of the pump.

Pump 1

Why do you think so? What about that pump made it better for pumping ketchup?

When you added Some water it went very far it when 175 centimeters if we didn't put a little bit of water it would come out the pump when we saueeze it for the liquid.

What are some other things you learned from your experiment?

I learned that ketchup is part liquid and part Gel because ketchup is made of tomatoes and a tomatoe is made out of a plant and the plant needs water to grow. And I learned that pump 1 is the best for all the liauids wedid.

November 6

I wanted the groups to present their results to the class using their graphs. I felt that they should be able to explain which pump worked best for each liquid, using supporting evidence from their test data. I hoped that the class as a whole would discuss why a given pump might have worked better. I also expected them to raise some further questions that had come up during the activity.

Unfortunately, there was a lot of talking while kids were trying to present. No one really offered any valuable comments until I took over at the end and started asking questions about the different graphs. One student did have a question.

RACHEL:

How could we decide which pump was best, since we got different results for each test?

The class decided that the spray pump was the best pump, but according to Ernie, sometimes we don't want the "best" pump—for example, in the case of ketchup. Some further questions that came up were:

- Does the size of the tube make it work better?
- Would the lotion pumps work better for thick liquids if they had a bigger tube?
- Why did we have to put water through the spray pumps first to get the ketchup to go through?
- Are there other gadgets for liquids that we could compare with these pumps?

The students reached some very insightful conclusions from this project. They recognized that a technology that is best for one application, such as spraying water, might not be best for another, such as dispensing ketchup. "Which is the best pump?" is a question that cannot be answered directly,

because it depends on what the pump is to be used for. Their questions suggest that they were also beginning to grapple with the concept of viscosity ("thickness" of a liquid), and its relationship to the diameter of the tube.

November 9

I made up an evaluation sheet for the students to fill out. (See Figure 4-31.) The evaluations went well. Kids were able to find their group weaknesses and also the most positive parts of their work. In general, everyone gave themselves and their peers high grades. I was happy to see that most everyone enjoyed the project.

Christine also offered her own evaluation of the project. She felt that it had succeeded in its primary goal of giving her students an opportunity to design and conduct their own controlled experiments:

As crazy as this activity was at times, I would do it again because the students really walked away having learned a lot. It was an excellent way to reinforce scientific methods, because they were allowed to design their own experiments. It was a great way to introduce the idea of product testing to the class, and get them thinking about creating fair or valid experiments. They were also able to practice metric measurement skills, which we had been working on in class.

Even though I provided a great deal of guidance, the students had a definite sense of ownership of their work during this project. They took an idea, created an experiment, and collected a data set on their own. They used the skills they had been working on in math and science, but this time with their own data.

4-31: **Maritza's evaluation form**

Pump Lab Evaluation

A. On a scale of 1 to 10, What is your score? __10__
   Give two reasons why you deserve your score.

   1. BEcause I coooperated with my group and behaved good with them

   2. I did almost all the work.

B. List each group member and give them a score between 1 and 10.

   1. Norman   Reason for score: 5 because he was talking alot.

   2. Claudette   Reason for score: 3 She didn't do alot of work and talked to much instead of paying attention

   3. Manuel   Reason for score: 10 because he helped alot with the group and always was paying attention

C. What did you enjoy about this project? That we got to do alot of things I liked it cause we played with water and ketchup, oil, milk, and soap I really liked this project cause it's alot of fun I hope we get to do it again

D. What could you have done differently on this project? I could of present it to the class we could of done different things with the group if they didn't talk so much with each other

Designing and Making Structures to Solve Classroom Problems

Both of the teachers featured in this section work in the upper elementary grades. These stories begin with a familiar classroom problem: lack of adequate storage space. In both cases, the students designed and built their own structures to solve this problem. Their primary material was cardboard from discarded packaging.

The Portable Storage System
by Sandra Skea

Getting Started

Students of Class 604 were creating dioramas based on scenes from the play "The Phantom Tollbooth." Clean-up time was especially difficult, for classroom storage space was severely limited and student lockers were already packed to capacity. The problem was complicated by the fact that we use several different rooms during the week.

Together, the class and I decided that if we were to continue a project-based exploration of mathematics, we would need to find an effective solution for our storage problem. We also realized that our solution must be a portable one in order to satisfy the constraints of moving from room to room.

Our goal became exploring how to build portable storage systems that would meet our requirements. We explored the problem and brainstormed options. We discussed storage units in general, and the classroom and school constraints in particular. We also used this time to conduct a scavenger hunt to search for materials for use in the construction of these systems.

After experimenting with the materials and with design possibilities the students developed criteria for the system. The criteria included the following:

- The system must be free standing and portable.
- The system must hold and support 8-10 shoeboxes.
- Each compartment must be uniform in size, and be able to support a live load of one pound.
- Accurate measurements are to be taken and recorded, including length, height, width, girth (measured two ways), face diagonals and volume.
- Recyclable materials are to be used.

Sandra Skea teaches math at Mott Hall, a middle school for high-performing students in Washington Heights, New York City. Sandra wanted to involve her sixth-grade class in a structures project that would both solve a real problem, and also require them to use geometry and measurement in a practical context. A real problem had already presented itself. The class had created shoebox dioramas, as part of an extended math project. However, they had nowhere to store them. Worse yet, due to a shortage of space in the school, the same class met with Sandra in several different rooms during the course of a week, and they needed access to these dioramas at each location. How could they both store them and also transport them reliably? Sandra explains what they did.

Sandra's class worked on this project for almost two months, devoting one-hour periods to it once or twice a week. Within the criteria listed above, most of the design decisions were up to the students. They were free to choose the recycled materials, design the geometry of the unit, add handles and/or doors, choose their own methods for attaching the parts, and develop their own methods for testing the design. Sandra also encouraged them to think about storage problems at home, and think about how similar storage units might solve them.

The students divided themselves up into groups of four or five to work on this project. Sandra required them to assign a specific job to each member of the group. Here is Sandra's description of the activities of the groups:

Design and Construction

Students talked freely during this activity. They engaged in sharing ideas, in questioning group decisions and in offering comments and suggestions. Students were allowed to circulate during the designing and building stages to ask other groups for tools and materials. Some questions and statements I overheard were:

Should we use inches or centimeters?

Which is better to use: masking tape or packing tape?

How can we make the shelves stronger?

Why are the shelves buckling? What can we do to fix it?

Where should we place the handles?

Are all shoeboxes the same size?

To make the box stronger, should we put tape on the edges?

Let's add a little space around each compartment so it will be easier for someone to pull out the dioramas.

Many of the design decisions and discoveries were interesting. Stephanie, Rosa, and Flor decided not to make a back for their box. Rosa said this would make the unit lighter and easier to carry. Flor added that it would also make it possible to see in from either the front or the back. (See Figure 4-32.)

Melida, Clarissa, Erica, and Alyssa recognized that students might use different-sized shoeboxes for their dioramas. How could they plan for that? (See Figure 4-33.)

4-32: **Stephanie, Rosa, and Flor decide their unit doesn't need a back.**

4-33: **Melida, Clarissa, Erica, and Alyssa examine shoeboxes of different sizes.**

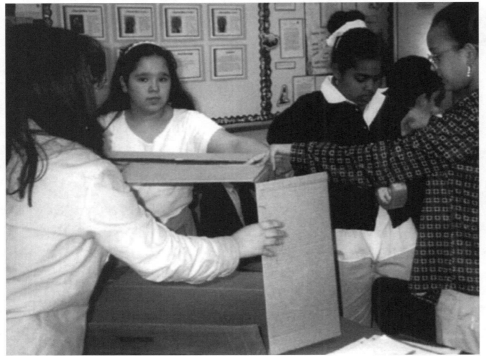

I was very surprised to see how closely students monitored each other's contributions. The atmosphere was one of dedication to task. Fooling around or time off-task simply were not allowed.

Sandra believes very strongly in ongoing assessment of her students' work. She used an impressive array of assessment tools and techniques, so that she and her students could monitor the progress of the project on a very frequent basis. Her methods included daily sharing sessions at the end of each class, regular homework assignments related to the project, final group presentations to the class, peer assessment of the presentations, individual written reports, and a self-assessment of each student's work. Sandra's reflections on assessment can be found in Chapter 5.

The students were given time to share their work at the close of each session. Each group assessed their progress and made plans for the next day. Questions and comments that arose during these sessions included:

What will happen if we double the cardboard?

What is the best placement for the handle(s)?

Should flaps be added so things don't fall out?

We all took turns and accomplished a lot. We cooperated with each other nicely.

My group calculated the area of the shoebox to see how much cardboard we will need.

Today I was the timekeeper for my group. We took measurements and got a plan ready for tomorrow.

Some of the homework assignments given in relation to this project included:

- Describe the plan you and your group developed to solve the storage problem of Class 604. Include the roles assigned to each member and how you decided the roles would be assigned.
- Describe ways in which your group could change the design of your portable storage unit to make it stronger.
- Describe and evaluate how well your group works as a whole.
- Describe how you might modify the design of your classroom portable storage system to create a structure that could be used to solve a storage problem unique to your household.
- Create a drawing of your portable storage system with all the dimensions labeled. Include length, height, width, girths, and face diagonals.

In addition to the regular homework assignments, every student had to prepare a written report about the portable storage system. It had to include how the structure was built, the dimensions of the system, how it might be improved upon, and other possible uses for it. Here are some of the ideas they had about applications for similar storage units at home:

MELIDA:
I would use this box to sort out papers that I will need later in the year. It would be like a little file cabinet, only homemade. In certain spaces I will put my laptop, papers, coupons, and receipts. My schoolwork will be a lot neater. I will find everything quickly. My mother will not scream her wits out because I decided to be irresponsible.

JENNIFER:
I would use it to keep my CDs in order.

ERIKA:
Instead of making a shoebox holder, I would make a towel holder. The unit can be mounted on the wall in the bathroom.

ALYSSA:
Our storage unit would also be useful in my apartment. I would use it to hold my personal belongings, such as my lip balm, camera, diary, Walkman, and portable CD player. I would make a couple (of) changes though. I would add more shelves, make it bigger and paint it light blue.

CLARISSA:
At home, I would use it as a bookshelf, because my desk is a mess, it makes my room look dirty. It would be great to keep all my books in one place.

ROSA:
Using a box design like this in my house would definitely be a helpful and handy way for me to keep my things organized. I can put my toys neatly into each compartment. I can also use it to store my nicely folded clothes.

Each group also had to make an oral presentation, in which they explained and demonstrated their design and answered questions about it.

These final presentations included the showing of their storage system, a description of how the group designed and built it, a demonstration of the mathematics involved, an analysis of the strengths and weakness of the system, and proposals for how the structure could be modified for use at home. During the presentation and the discussion periods that followed, students shared ideas and asked questions, and offered comments. (See Figure 4-34.)

Questions included:

What are the handles made of?

How else can you use your system?

How strong is it really?

How many shoeboxes can it hold?

Show me how you measured the girth.

LUIS:
Why did you put the handle on the top instead of the sides?

JULIAN:
That way, things will not fall out, because when we carry it, the back will be the bottom.

Comments and suggestions included:

You need to reinforce the handle. It will not stand up to wear and tear.

The compartments are of different sizes; perhaps they should be uniform.

You need to strengthen the support system for the shelves.

You used too much tape.

4-34: Storage unit with handles on top

Omar described some of the changes he and his group made to their structure. The top of the unit was originally a trapezoid, but this design wasted space and caused the compartments to shift. (See Figure 4-35.)

4-35: Omar explains how they changed their design.

Rachel, Precious, and Daniella used two modules to build their unit. They said that if they made another structure, they would use less tape. They also discussed using other joining methods, noting that nails would not work. Rachel noted, "I tried them and they fell out." (See Figure 4-36.)

Sandra prepared a peer evaluation form for students to fill out during the group presentations. (See Figure 4-37.) These forms, which were anonymous, provided each group with their fellow students' assessments of how well they had presented.

Each student also had to evaluate his or her own participation in the project, including both group and individual work, on a self-evaluation form. (See Figure 4-38.)

Candice was very self-critical:

If my group and I had planned more carefully, and worked more diligently, our box could have been much better. We should have recorded our measurements so we would have known how many shoeboxes could have fit into the storage unit. My group didn't plan time carefully, so we ran out of it.

Perhaps the most authentic form of assessment of a design project asks about the design itself. Did it really solve the problem it was intended for? What do the users think of it? A few months after the conclusion of the project, Sandra wrote:

4-36: **Rachel, Daniella, and Precious present their storage unit.**

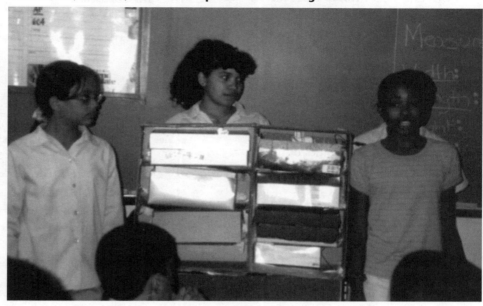

4-37: **Rating Guide for Group Presentations**

CITY COLEGE CURRICULUM GUIDES
98 Portfolios

The Portable Storage System

Rating Guide for Group Presentation

Directions:
    Listen carefully as each group presents. Please do not interrupt the speakers. You will have time at the end of the presentation to ask questions or to make comments. Rate your classmates fairly by circling a number from 1 to 5. A rating of 1 is low; the highest rating possible is 5. Please be sure to read the criteria carefully. Use the back of the rating sheet to record any additional comments or suggestions you have.

CRITERIA		RATING			
1. Students spoke clearly, and used complete sentences.	1	2	3	4	5
2. Measurements were taken and recorded accurately.	1	2	3	4	5
3. Students described how they built their structure.	1	2	3	4	5
4. Students explain how they could change the structure to make it stronger	1	2	3	4	5
5. Students suggested a way they could use the structure at home.	1	2	3	4	5
6. Every member of the group participated in the presentation.	1	2	3	4	5
7. The storage unit is free standing and portable.	1	2	3	4	5
8. The unit has 8-10 compartments capable of holding a shoebox	1	2	3	4	5
9. The storage unit is attractive.	1	2	3	4	5
10. The unit uses space efficiently	1	2	3	4	5

I was also delighted to discover that the storage system created by the students actually did solve our classroom storage problem. As a matter of fact these systems are still in use. At present, they house the materials for our next project.

Overall, Sandra was very pleased with the educational outcomes of this project. It served to develop some mathematics concepts in ways that traditional instruction does not.

I discovered the only real way to assist students in understanding some concepts is through a hands-on experience. Examples are the relationship between length, height, and width to volume, the relationship between girth and perimeter, and the process for measuring and calculating face diagonals

The study of geometry, rectangular prisms in particular, became accessible and real. Students who previously had difficulty calculating girth, volume, and face diagonals from two-dimensional drawings gained the confidence and the skill to do so with ease and accuracy. The understanding of girth, the two ways to measure it, and its relationship to perimeter, became clear as students used their own three-dimensional rectangular prisms for exploration of mathematics.

Building a rectangular prism also served to improve and extend the students' understanding of dimensions and of measurement. Growth also occurred in their ability to use and compare measurement tools. And it must be noted that attention to homework increased and test scores improved.

4-38: Clarissa's self-evaluation

CITY TECHNOLOGY CURRICULUM GUIDES

Student Self Evaluation

1. The part of the project I enjoyed the most was _the finishing touches_ because _the group really connected at that moment we worked together_

2. The hardest part of the project was _all the measuring parts_ because _sometimes we would fight about who was right._

3. I helped the project succeed by _I cutting the cardboard almost at an exact length and bringing in materials_

4. The math that helped me to enjoy and to understand this project was _Geometry, measurements of face diagonals and corners_

5. The written work I did related to this project was _a report about how to make it better, how would you use this at home. also note copying._

6. The part I like writing about best was _the report and the drawings_

7. I can use what I learned from this lesson when I or if I _build a structure at home or other subject areas_

8. On a scale from 1 to ten I rate this project a (an) _9+_

Use this space to record any additional comments you would like to make.

I like doing this project. I think it help understand geometry and measurements.

I enjoyed working with my class who encouraged me to do better.

I would'nt give this a 10 because it deserve a 9 ½ because at some points they were tough decisions to make. And sometime it became very boring. IF it were not because of this I'd give it a 10 plus

Two students in particular became very excited about attending math class. Previously, they had contributed little to classroom discussion, rarely completed assignments and class work, and in general saw no reason for studying mathematics. Grades went up for both students.

Although Sandra's primary goals are in the area of mathematics, she saw some other benefits from this project as well. Her students had learned about the value of recycled materials, and had gained some experience in designing and constructing useful structures from them. They discovered that a group is greater than the sum of its parts, because they saw how group members could stimulate and reinforce one another's ideas. They also gained experience in both written and oral presentation. Perhaps most important, Sandra felt they had discovered that they had the power to improve their own environment.

Students explored how purpose affects the design of a structure. They learned how to locate and use recyclable materials in class projects.

They learned the advantage of staying on task and of sharing job responsibilities. Students discovered there are many ways to solve a problem and that brainstorming and the sharing of ideas can simplify the process.

Students also learned how to write clear, concise, and complete reports. The presentation component gave them the opportunity to develop and enhance their listening, speaking and questioning skills.

Students were also involved in an experience where they were in control. The activity was a project for investigation and exploration, not just another routine assignment. Mathematics and technology can be used creatively to solve problems. The students learned that they can have power over their surroundings.

Learning About Structures, Preparing for the Science Fair, and Solving a Storage Problem *by Michael Gatton*

Michael Gatton taught science to sixth- through eighth-grade students at I.S. 143 in Washington Heights, New York City. He was an experienced science teacher who was required to use some materials from Insights, *a science curriculum package. He was also responsible for preparing his students for the district Science Fair.*

Michael developed and implemented this unit after doing the "Structures" module from *Insights*. He involved his students in designing and building a badly needed storage unit for their book bags. In preparing for the design of this storage unit, each of his six groups did a controlled experiment that served as the basis for a Science Fair project. The results of three of these experiments provided data that the students used in their storage unit designs. As a whole, this project offers a model for integrating scientific analysis with technological design. Michael's story begins with the logistics of his sixth-grade class, his selection of material, and the requirements of the sixth-grade science curriculum.

The Problem

The sixth grade in my school doesn't fit neatly in either the "elementary" school or the "middle" school category. The students have an official classroom where most of their subjects are taught, which is the elementary school model. On the other hand, each subject is taught by a different content area teacher, which follows the middle or junior high school model. They do travel to different rooms for certain subjects, but this is due more to the overcrowded conditions in our school than to some logical scheme. Other teachers and other students use our room throughout the day, and my students must periodically pack up and move to another room.

Some major problems result from the multiple use of the room. If my students leave their materials in the classroom, they could be stolen or "borrowed" by other students who use our room. There is sometimes an adversarial relationship between my students as they put things away before leaving the classroom and the students and/or teachers who are trying to get into our room for the next lesson. Complicating matters is the school policy that closets are to be locked in the morning and not opened until dismissal. These closets are supposed to be used exclusively for jackets and caps. Of course, students frequently leave books or papers in the closet, and have to interrupt another class to retrieve them.

Clearly, what is needed is a storage system that offers both easy access to the students as they exit or have to retrieve something, and security so that their things will not be taken. With these two goals in mind we set out to design a structure that would fulfill those needs.

Material

What material should we use to make our storage structure? We decided to use cardboard, because it is easily available and free of cost.

However, there are a number of characteristics of corrugated cardboard that make this material difficult to use. It is not easy to cut straight without a box cutter or similar sharp blade. It warps when wet with glue. It has to be recycled in the school, which creates a lot of extra work for the custodian in terms of stacking and tying the odd-shaped scraps. It can be quite strong when folded into the box configuration but individual pieces bend quite easily. It is not so easy to find large pieces for a full-sized unit such as we designed. These were problems we would have to address together.

Sixth Grade Science

Our sixth-grade science curriculum consists of three kits designed by *Insights* and covering three different areas: earth science, physical science, and life science. Our module for physical science is called "Structures." The unit focuses on some key concepts in structural design such as structural components, loads, materials, and shapes. In preparation for this project we did extensive work in the *Insights* curriculum on the concepts mentioned above. A weakness of this unit, however, is its reliance on models of big structures such as buildings and bridges. Students do not have an opportunity to design a real-world structure for their own use. I decided to add this element to the structures curriculum.

An additional key feature of our science program is the annual I.S. 143 Science Fair, which is fairly traditional. Students are expected to conduct and present a controlled experiment that is related to some part of their science curriculum. The design challenge of building a strong cardboard structure presented a wealth of possibilities for Science Fair projects. Here are the kinds of questions we could try to answer through controlled experiments:

- Which cardboard should we use?
- How could we laminate the cardboard to make it stronger?
- Which type of glue would work best?
- How should we attach the shelves?
- What shapes should we use for the columns and beams?

The Design Challenge

To motivate the students I began by discussing their frustrations in moving from room to room. On average, this movement occurs three times per day, including lunchtime. Typically, my students would travel to another room for computer lab, for reading lab, for gym or for social studies. At the end of the day there is usually an after-school program that uses our room and makes it impossible for students to leave materials in their desks.

I asked if they were getting tired of carrying their books around all day, and of coming in every day to find paper or pens missing from their desks. I informed them that if they were interested, we could design and build some storage units to hold their belongings and make life a little easier for them. They definitely liked the idea.

The challenge, then, was to design a system of storage units out of cardboard that would be strong enough to support the load of some full book bags and a few other items, provide a certain amount of security, be easily accessed and not be an eyesore.

Establishing Criteria

We held a brainstorming session about criteria. The most important criteria that we came up with were as follows:

- Each compartment should hold the size and weight of a full book bag. A book bag could weigh as much as 20 lbs., and be as big as 60 cm. x 45 cm. x 30 cm.

- There should be six compartments per unit.
- Doors of some kind should be provided for security.
- They should permit easy access.
- Their locations in the room should allow traffic flow.
- They should look nice and neat.

We decided to keep the bottom compartment empty so that no one would have to get down on the floor to get his or her stuff. The size of each compartment was dictated by the size of a large book bag.

In deciding on security, we all realized that cardboard would be fairly easy to break into if anyone really wanted to. On the other hand, two of the storage units used to hold my personal materials were never locked and nobody had bothered them all year long. It seemed that theft would not be an issue unless the items were out in the open and easy to get to. So we thought that a simple door with some kind of latch would work fine for our needs.

Designing the Storage Units

After we came up with the criteria and the material, our design options were fairly limited. If you want a structure of certain dimensions built of cardboard and having a certain number of storage spaces and offering a certain amount of security, you are pretty much locked into a basic shelf unit design. Nonetheless, I didn't tell students this. I simply asked them to design a structure on paper, which would fit our criteria. They all looked pretty much the same: a basic enclosed rectangular shelf unit. (See Figure 4-39.)

4-39: **Design of shelf unit**

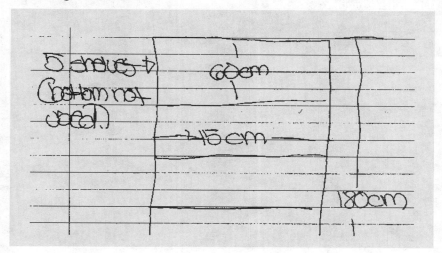

The next step would be to make the drawings to scale. The goal was to create a scale model so that the students would see some of the problems we would face. Then we could come up with some strategies for creating the real thing. I had to make the decision, in the interest of time, to simply give students the scaled down measurements. I have taught scaling to another class, in the context of scale maps, and it is a difficult concept for them. Teaching students how to scale might have taken another week or two in itself. Instead, I demonstrated for them what we were doing when we scaled down, but I did not expect them to be able to reproduce what I had done.

Making Scale Models

From the basic design we discussed the idea of putting together a scale model. That way, we could all see in 3D what our shelf units would look like and get some sense of how to put the units together.

I was surprised that the students had no idea how to go from the scale drawing to a scale model. I had thought that by discussing the sizes of the various parts of the unit, it would be obvious how to measure the pieces and cut them out. To my surprise, some of the groups started randomly measuring pieces of cardboard with no regard for the scale drawings. Others ignored the scale, abandoned all attempts at measuring and seemed to be trying to mold the entire unit out of a single piece of cardboard. I was beginning to realize what a complete lack of experience my students have in building things, or in understanding how things are built.

As teachers often do, Michael had taken it for granted that his students would know some basic procedures that turned out to be entirely new to them. He backtracked and did a bit of "reverse engineering." He had them look at manufactured items around the

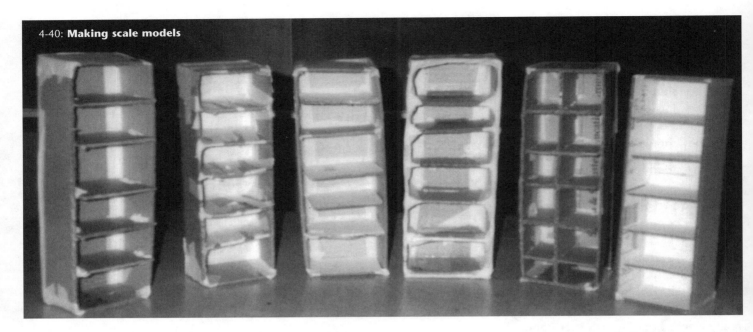

4-40: **Making scale models**

classroom, as examples of how structures are put together. In looking for structures to study, one never has to go very far!

I stopped the lesson and went around the room pointing to structures and discussing how they are manufactured from parts that are assembled to make a whole. We discussed each of the parts that were needed to build our model shelf units. We wrote down both the dimensions and the number of parts required. For example, we needed two side pieces that would be 27.7 cm. x 6.9 cm. I then discovered that they didn't know where to begin with creating a rectangle. I had to demonstrate how to measure and draw a rectangle.

After we completed cutting out our pieces, we assembled our scale-model units and discussed some of the problems that we would face when building the real thing. (See Figure 4-40.) For starters, our model units were fairly strong due to the sizes of the pieces. When we were to

build a larger unit, however, the cardboard would not maintain the same relative strength. This meant we would need some way of laminating pieces of cardboard together to make stronger, thicker pieces.

We needed some way of supporting the shelves. We also thought about how we might incorporate some supporting columns and how we might change the shape of the shelves to make them stronger. We decided that tape would make for an ugly structure, and decided that we would use only glue and cardboard.

In making and thinking about the scale models, a number of issues came up that could only be resolved by some systematic testing. These tests would also be the basis for Science Fair projects. But first, the students would need to understand the need for controlling variables, and develop the concept of a "fair test." Michael recognizes that students are unlikely to learn these principles just by being told to follow them. They have to see the need for themselves.

Testing Variables

The problems that we identified in building models led us to test some variables in order to discover the best design for our shelves. This is where Science Fair projects naturally fit in. Students were assigned to groups and each group was assigned one of the following research questions:

1. How do different types of glue affect the strength of a laminated shelf?

2. How do different ways of laminating cardboard affect the strength of a shelf? For example, should the ribs of the two pieces of cardboard be oriented the long way, the short way, or some of each? (See Figure 4-41.)

3. How do different types of cardboard affect the strength of a shelf?

4. How do different shapes affect the strength of a shelf?

5. How do different types of triangles affect the strength of a truss?

6. How do different types of supports affect the strength of a shelf?

The concepts of a controlled experiment are not easy to grasp and instruction in this area is an ongoing process. I had already done an experiment connected to the *Insights* module as a neat way of teaching the basic elements of a controlled experiment. (See "How Does the Shape of a Column Affect Its Strength?" in Chapter 3, "Activities.") These concepts include independent and dependent variables, constants, data collection, and so on.

I usually try to give my students a concrete experience first which I hope demonstrates the need for certain vocabulary. So we build vocabulary on top of the experience and I can always refer back to the experience for review. Because I teach most of my students for three years, I planned to spend a great deal of time with them in sixth grade learning how to do a Science Fair experiment. I expect that they will do more and more of their own work in the seventh and eighth grades.

Before students began their Science Fair projects, I again reviewed the requirements for a fair test. In response to my review questions they seemed to get most of the concepts, at least in the abstract. In practice, however, students had a hard time identifying variables in their own tests and needed a lot of prodding to keep all the variables constant except for their independent variable.

Although I repeatedly reminded students about controlling variables, and they kept reassuring me that they had done so, I deliberately allowed them to make mistakes. For example, a group of students con-

4-41: **Testing lamination methods**

ducted an experiment on which type of glue would make the strongest laminated shelf. With my guidance they kept the size of their pieces constant and then glued the pieces together to form laminated beams. They tested the load capacity and declared that Elmer's School Glue was the strongest. "Great," I said. "How much of each type glue did you use?" "Uh-oh. We didn't measure the amount of glue."

So they had to redo the experiment keeping the amount of glue constant. Every group had at least one obvious variable that they initially neglected to control. In each case we discussed the problem as a class in terms of what constitutes a fair test. It is a time-consuming approach, but it's really the only way students will grasp the idea of controlling variables.

When the Science Fair projects were finished, we discussed how to incorporate the findings into the construction of our shelf units. In some cases the results were inconclusive; in others, we simply decided not to use the particular variable. For example, the shape of the shelf couldn't really be changed, and it would have been impractical to use trusses to support the shelves.

While some of the experiments didn't turn out to be so useful to the design project, others were. Here is an excellent example of how science inquiry, using a controlled experiment, can inform a technological design. The results of the experiments on glue, lamination method, and support method really contributed valuable information:

1. A nice outcome of the glue experiment was that the most readily available glue turned out to be the strongest: Elmer's School Glue.

2. The experiment on laminating cardboard pieces was very instructive. The laminate in which all the "ribs" are running the same way (along the length, not the width, of the beam) was the strongest combination for making a shelf or beam.

3. Also very useful was the experiment on which supports work best for holding up a shelf. Here we found that a simple strip of cardboard, glued to the back and sides of the storage unit, provided the most support.

Final Product

With the results of our tests in hand, and with time rapidly slipping away, we decided to create a single shelf unit. Each group produced a laminated shelf. The first groups finished were assigned to laminate the side and back pieces and make the support strips. Then, just as we had done with the model units, we assembled the pieces to make a whole unit.

Of course, with such a large piece, the assembly part proved a bit more involved and required several stages. As this was our first unit, we weren't quite sure what order to follow, and there could easily have been a better way. If we had time to produce a second unit, we would definitely do some things differently. Nonetheless, here's how we did it:

1. Lay the back and side pieces flat on a table.

2. Measure and draw lines where the support strips will be glued.

3. Glue the support strips onto the sides and backs.

4. When the support pieces are dry, glue the edges of the back and the sides where they come together, place the shelves on the support pieces (without glue), and fold the side pieces up and into their places. The shelves are there to keep the sides in place.

5. Use twine to hold the glued sections together.

6. Place the shelf unit on the floor, on its side, and use weights to hold the pieces in place. (See Figure 4-42.) Allow it to dry. (The result is shown in Figure 4-43.)

7. Glue the shelves to the support strips. (See Figure 4-44.)

8. Make cardboard flaps and glue them to the outside corners to stabilize the unit. (See Figure 4-45.)

Ideally we would have attached doors to the unit. We would then have tested the unit for strength and made adjustments to the design. We would then use the experience of building a single unit to construct five more units. Our plan proved overly ambitious and I will suggest ways to avoid this problem.

4-42: **Gluing the sides to the back, using the shelves to keep things in place**

Overall, Michael was pleased with the outcome of the project. Besides producing a useful piece of furniture and six Science Fair projects, it had engaged his students in learning math and science in the context of a real problem and given them a genuine feeling of accomplishment. At the same time, there had been many frustrations along the way, and the project had taken considerably longer than he had expected.

These difficulties were mostly inevitable, because of the way in which a design and construction project draws upon such a wide range of skills and concepts that students may not have. As Michael points out at the end of he following passage, many skills and concepts that are taught in math or science are never really learned until students really have to apply them.

4-43: **Back and sides, showing support strips**

4-44: Gluing shelf to support strip

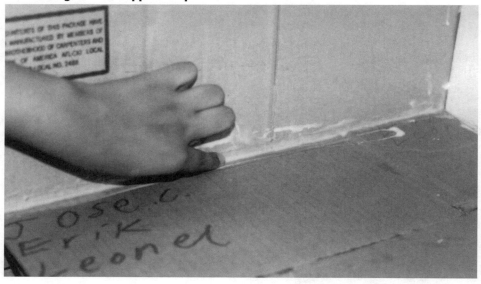

4-45: Adding cardboard flap to strengthen back-to-side joint

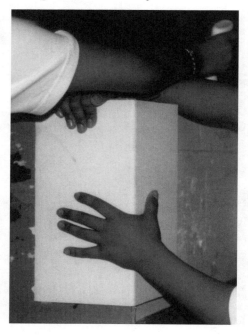

Frustrations

After we completed the scale drawings, we ran into a bit of a snag that any teacher who wishes to do this project must address. How does one cut the cardboard? Here's the dilemma: Box cutters are neat and efficient, producing nice clean lines rather quickly, but they are dangerous and are not even allowed in many school buildings. As I discovered, you might also get tired of cutting all those pieces of cardboard yourself. On the other hand, any other method that we tried proved slow and produced rough, aesthetically unpleasant lines that might prove difficult to glue.

Another problem came with the actual construction of the scale models. Perhaps I should not have been surprised that students lacked the basic understanding of how things are put together in the real world. Many of them wanted to mold the entire model out of a single piece of cardboard and most of them completely ignored the dimensions that we had agreed on. This was of course a frustrating realization and one of many such construction problems that extended this project beyond the original timeline.

Each task that I wanted students to learn had at least one subtask that they needed to learn before doing it. So in the example above, I wanted students to learn about controlling variables, but I had to help them measure the mass of a liquid in order to do this. It's a frustrating realization that students often do not come to us with the prerequisite knowledge that we expect them to have. Then it becomes difficult not to get caught in an almost infinite regression teaching what should be prior knowledge. How many times have I used science class to review or even teach basic multiplication and division because a science activity requires students to perform averages?

I will provide the following list of the necessary concepts and skills that many of my students simply didn't have, as I often discovered the hard way:

- Measuring: using a ruler, scale, graduated cylinder, triple beam balance
- Right angles: for making good rectangles, shelves, columns, etc.
- Place value: for reading a metric ruler and working with measurements
- Making data charts and graphs
- Multiplication, division, and taking averages
- Making parallel and perpendicular lines

In many cases, I suspect the problem is simply one of applying to a real-world task the skills and concepts that have been taught as abstractions in math or previous science classes. That's what makes these types of projects so important, and why I consider the time spent on this project worthwhile. It brings math, science, and technology together.

Streamlining the Project

How could someone do this project without spending almost half a year on it? I believe that the idea of building single large units does not work well. It requires too much teacher coordination and perhaps too much material. If I were doing this project over again, I would have each student build his or her own individual compartment and make them stackable. This would eliminate the need for large pieces of cardboard and would make lamination a lot simpler. It would also tend to keep everyone engaged and invested in the project.

The potential problems I see for this approach would be getting the units to a standardized size and construction. There will always be some students who are less adept at making things, and this will create problems with lining up the units and stacking them. I would also find a way to let students do their own cutting. There may be other ways of cutting that we didn't consider. There may be a way to make the edges look nice after they have been cut. It might be possible to use papier mache or some other technique to cover the cardboard and hide the rough edges.

Evaluation of the Project

I was impressed with the final product. It was a sturdy piece of furniture with a lot of potential. There were mistakes made in the construction that would have been simple to correct if we had continued. For example, the students didn't always follow the rules for laminating that we had agreed upon. It showed in the strength of some shelves. These shelves would need to be stronger or they would need more support in the front.

I found the experience of building a shelf unit out of cardboard to be both rewarding and frustrating. It's rewarding to see certain students who don't necessarily shine in an ordinary classroom environment really come through and perform enthusiastically in this type of activity.

One of my students, for example, is diagnosed with ADD and normally has trouble staying focused in class. He probably worked harder than anyone else on his part of the project and produced a Science Fair project almost single-handedly (the other members of his group weren't as enthusiastic). I also believe that my students come away with a pretty good idea of how to do a controlled experiment, which I consider an essential part of science education.

I believe that linking the Science Fair projects to a design challenge is an innovation that we will add to the school-wide Science Fair guidelines next year.

Chapter 5

RESOURCES
Help for Teachers

Making Connections with Literature

Using literature as a supplement and enhancement for instruction is good teaching practice because:

- Children learn from everything they experience.
- Children learn more effectively when instruction is associated with positive emotions, such as those evoked by a good book.
- Literacy is key to children's success as learners.
- There are many different learning styles.

We encourage you to incorporate books of all kinds into your work with *Packaging and Other Structures.* We've included an annotated list of quality books of all kinds on the following pages—storybooks in which the uses and design of packaging and other structures are demonstrated, as well as nonfiction books on how to design and create packaging and structures, the functions they serve, and how to make the best use of materials.

But don't stop with these. You know your students and how they learn better than anyone else. When you see a book that might further your instructional goals, interest or challenge a particular student, or evoke feelings that make learning more fun, add it to the books that are available to your students.

Amelia's Road, by Linda Jacob Altman. Lee & Low Books, Inc.: New York, 1995. (Recommended grades: K-5)

Tired of moving around so much, Amelia, the daughter of migrant farm workers, dreams of a stable home. She finds peace at the story's end by planting a treasure box, which is filled with her treasures, beneath a wondrous tree. This book allows older readers to see that families live in all kinds of ways. Her treasure box packages all of Amelia's dreams.

And So They Build, by Bert Kitchen. Candlewick Press: Cambridge, MA, 1995. (Recommended grades: PreK-3)

Students can explore the extraordinary world of animal behavior with descriptions of 12 astonishing animal architects and explanations of why and how they build their marvelous structures. Color illustrations show animals, especially insects, as they work building their structures.

Ant Cities, by Arthur Dorros. HarperCollins: New York, 1988. (Recommended grades: K-4)

A "Let's Read-and-Find-Out" science book that explains how ants live and work together to build and maintain their cities. This literary connection presents the structural problems ants contend with and explores their cooperative working habits.

Art from Packaging, by Gillian Chapman. Raintree Steck-Vaughn: New York, 1997. (Recommended grades: 2-6)

Contains instructions for making prints from bubble-wrap, puppets from cardboard boxes, and other arts-and-crafts projects using discarded packaging materials.

Bad Egg: The True Story of Humpty Dumpty, by Sarah Hayes. Little Brown and Company: Boston, 1987. (Recommended grades: K-2)

Humpty Dumpty sat on the wall. Humpty Dumpty had a great fall. What events happened in between? This true story of Humpty Dumpty, who it turns out, was not entirely blameless. An interesting link to use for introducing packaging to younger students.

The Bag I'm Taking to Grandma's, by Shirley Neitzel. William Morrow & Co.: New York, 1998. (Recommended grades: K-3)

A young boy is packing a bag for a trip to visit his grandmother. He fills a shopping bag with his mitt, cars, space ship, wooden animals, his favorite stuffed rabbit, his pillow, a book, and a flashlight. Then along comes Mom with packing ideas of her own. This literature link utilizes pictures, repetitive phrases, and rebuses to increase the fun for beginning readers, while integrating the theme of packaging.

The Big Box, by Virginia Clammer. Millbrook Press: Brookfield, CT, 1999. (Recommended grades: K-2)

Bill and his little sister Kay enjoy playing in a big box, which becomes in turn a car, a jet, and the engine of a train. This "Level 1 Real Kids Reader" blends phonics with whole language. A beginning controlled vocabulary helps connect packaging with literacy for emergent readers.

A Box Can Be Many Things, by Dana Meachen Rau. Children's Press: New York, 1997. (Recommended grades: K-3)

After their mother tosses out a large box of junk, a girl and her younger brother retrieve it from the garbage. They pretend it is a cave, a car, a house, and a cage. Even when it lies in pieces on the ground, their imaginations lead them to more inventive uses for the box.

Boxes! Boxes!, by Leonard Everett Fisher. The Viking Press: New York, 1984. (Recommended grades: PreK-2)

Enticing rhymes explore all the things a box can be—box-kite, jack-in-the-box, candy box, treasure chest and more. Colorful illustrations transform everyday objects and expand the playtime of young readers and listeners.

The Bridge, by Emily C. Neville. Harper & Row: New York, 1988. (Recommended grades: K-3)

When the old wooden bridge breaks, a young boy is delighted to be able to watch, from his front yard, the many different machines at work building the new bridge across the brook. The new bridge is a dirt road over a culvert. The new structure is more sturdy, but not as unique and memorable as the old rattling wooden bridge.

Bridges! Amazing Structures to Design, Build and Test, by Carol A. Johmann. Williamson Publishing Company: Charlotte, VT, 1999. (Recommended grades: 3 and up)

Each chapter consists of numerous short articles combining historical and technical information on the design and construction of bridges with easy hands-on experiments. It includes analysis of arch, beam, and suspension systems, the "care and feeding" of structures and reflections on bridges of the future. Throughout, there are simple projects involving building, measurement and observation as well as writing poems about bridges. The book concludes with a challenge for young minds to "think outside the box."

Charlotte's Web, by E.B. White. HarperCollins: New York, 1976. (Recommended grades: 3-6)

"Some Pig!" These words that Charlotte spins into her web to describe Wilbur cause plenty of excitement, and ultimately, help save his life. A classic that depicts life and death, the passage of time, and the wonders of nature, and also shows readers that true friendship lasts forever.

Creative Crafts from Cardboard Boxes, by Nikki Conner. Copper Beech Books: Brookfield, CT, 1996. (Recommended grades: K-3)

Crafts are easily created from boxes and transformed into a train, car, ship, or puppet stage with puppets. The same supplies can create a game, a snake, or a wagon from paper cups. The book has no text, only numbered visuals, so no reading is involved.

Creative Crafts from Plastic Bottles, by Nikki Conner. Millbrook Press: New York, 1997. (Recommended grades: K-3)

Provides instructions for making such items as bangles, a rocket, a buggy, and a plant pot from plastic bottles. Wordless instructions and visuals explain each project.

Creative Crafts from Plastic Cups, by Nikki Conner. Millbrook Press: New York, 1996. (Recommended grades: K-3)

Provides a variety of craft projects for young children using plastic cups. There is no text—only numbered, step by step, visual instructions.

Dinosaurs to the Rescue, by Laurie Brown. Marc Brown. Little, Brown & Company: Boston, 1994. (Recommended grades: 1-6)

A guide to protecting our planet. Packed with good advice on how to use less of the earth's precious resources and how to find new uses for old household items. Text and illustrations of dinosaur characters introduce the earth's major environmental problems and suggest ways children can help.

The Great Trash Bash, by Loreen Leedy. Holiday House: New York, 1991. (Recommended grades: 1-5)

The animal citizens of Beaston discover better ways to recycle and control their trash. Young readers learn how to dispose of packaging materials and bash the trash in their own communities.

Houses of Adobe: The Southwest, by Bonnie Shemie. Tundra Books: Plattsburgh, NY, 1995. (Recommended grades: 4-6)

An excellent resource on the adobe structures built by Native Americans in the southwestern region of the United States. It describes the materials used, construction methods, and the uses of these different structures. Good link for integrating geography and packaging, and also a resource for student reports.

How Do They Package It?, by George Sullivan. Westminster Press: Philadelphia, 1976. (Recommended grades: 6+ and Teacher Reference)

Everything you always wanted to know about the world of packaging, as tubes, boxes, bottles, flip-top cans, aerosols, and jars compete for places on supermarket shelves. Different types of convenient, commanding, and colorful packages are described. Discover who created the first tea bag, why we buy milk in cartons and plastic containers. and why they don't package peanut butter in a tube. Though slightly dated, it provides fascinating facts and information for teacher reference and student research reports.

How Insects Build Their Amazing Houses, by W. Wright Robinson. Blackbirch Press: Woodbridge, CT, 1999. (Recommended grades: 5 and up)

Explores how various animals live and what can be learned from studying the types of structures they build. Describes step-by-step how termites build mounds; how wasps build nests; how ants build giant anthills; and how bees build nests, tunnels, and special chambers.

How on Earth Do We Recycle Plastic?, by Janet Potter D'Amato. Millbrook Press: New York, 1992. (Recommended grades: 4 and up)

Discusses the environmental problems caused by the manufacture and disposal of plastic, and presents craft projects that use recycled plastic.

Likeable Recyclables: Creative Ideas for Reusing Bags, Boxes, Cans, and Cartons, by Linda Schwartz. The Learning Works, Inc., 1994. (Recommended grades: 1-6)

Presents an endless array of fun-filled ways on how to decrease the garbage we produce from discarded bottles, boxes, cans, cartons, cups, and tubes, by making toys, games, and other crafts out of items we usually discard.

Little Bear's Trousers, by Jane Hissey. Philomel Books: New York, 1987. (Recommended grades: K-3)

While looking for his beloved red trousers, Little Bear discovers that other animals have found many uses for them: sails for a boat, a container for dog bones, a hump warmer or a hat, a flag, even a cake frosting bag. A very creative resource for the concepts of structure and function.

Look What You Can Make with Paper Bags, by Judy Burke. Boyds Mills Press: Honesdale, PA, 1999. (Recommended grades: K–6)

Each project in this book begins with a paper bag. Full-color photographs of the finished projects motivate children to explore and complete their own projects. From a "Rustic Log Cabin" to an "Autumn Scarecrow" to "Mask Mania," these appealing craft ideas will stimulate the imagination.

Lunch Boxes, by Fred Ehrlich. Penguin Putnam Books: New York, 1991. (Recommended grades: K-3)

The vocabulary is simple and repetitious, making this book appealing for young readers. It's lunchtime at Oak Hill School, and all the children go quietly to the lunchroom carrying their lunch boxes. Once they sit down to eat, their food goes everywhere but in their mouths. Various packaging items can be observed and identified on the lunch table when the lunch boxes are opened.

Maebelle's Suitcase, by Tricia Tusa. Macmillan Publishing: New York, 1999. (Recommended grades: K-3)

One hundred-and-eight-year-old Maebelle (the Pippi Longstockings of the geriatric set) designs and makes hats. She lives in a treetop house and has a lot of bird friends. When her neighbor flies over to borrow a suitcase to fly south, the packing creates a problem, but the resolution is warm and wonderful, just like the book.

Making Gift Boxes, by Linda Hendry. Kids Can Press: Buffalo, NY, 1999. (Recommended grades: 3-6)

Now kids can make their own gift boxes. Excellent step-by-step directions are given for projects that range from simple to challenging. An empty cracker box becomes a fierce-looking monster, or a small milk carton becomes a cozy cottage. Kids learn how to recycle easily found items into beautiful gift boxes. There is a color photo of each finished project.

Mommy's Briefcase, by Alice Low. Scholastic, Inc.: New York, 1995. (Recommended grades: PreK-2)

Look what's inside Mommy's briefcase! Packed inside this hands-on book are special folders containing all the things a working Mom carries in her briefcase. Neat link for introducing emergent readers to packaging.

The Mud Flat Mystery, by James Stevenson. Greenwillow Books: New York, 1997. (Recommended grades: K-3)

When a large box is delivered to Duncan while he is away, the other animal inhabitants of Mud Flat are consumed with curiosity about what might be inside. Is it heavy? Or fragile? Should they guard it? Is it food? Can they solve this mystery? An easy read-aloud book that will help kids along the road toward independent reading.

My Cat Likes to Hide in Boxes, by Eve Sutton. Scholastic, Inc.: New York, 1973. (Recommended grades: K-2)

Delightful book with rhymes about cats from all over the world and "my cat" that likes to hide in boxes! The predictable pattern encourages reading participation, either whole-group or independent. The idea of using boxes and shapes as homes is an early connection to structure and geometry.

Native Dwellings, by Bonnie Shemie. Tundra Books: Montreal, Canada, 1991. (Recommended grades: 2-6)

This is a series of books that includes Houses of Hide and Earth (Plains), Houses of Bark (Woodlands), and Houses of Snow, Skin and Bones, (Northern). Each volume deals with a type of Native American dwellings, including how they are lived in, as well as the building materials, techniques, and tools used to construct them.

Packaging Source Book, by Robert Opie. Chartwell Books: Secaucus, NJ, 1989. (Teacher Resource)

Pictorial guide to packaging design, from 1880 to the present. It chronicles the changes from the ornate and whimsical fancy goods wrapping of the 1880s to the bold packaging ideas that vie for consumers' attention today. This international survey of packaging highlights the innovations made in the design and packaging of everyday products, and the creation of brand images. An excellent resource for teacher reference, and an inquiry tool for independent student reports.

Plastics, by Terry Cash. Garrett Educational Corporation: Ada, OK, 1990 (Recommended grades: 3-6)

This book introduces plastics, showing how everyday things, from bags to tubes, to toys, are made and recycled, and what properties are characteristic of each material. It includes ideas for a variety of simple projects and experiments. An outstanding feature is the integration of questions and suggestions for collecting, observing, experimenting, and comparing.

Plastics and Polymers, by Robert C. Mebane. Twenty-First Century Books: New York, 1995. (Recommended grades: 6-up)

Explores the basic properties of metals, plastics, and polymers. Contains simple demonstrations that require little preparation and utilize easily obtained materials. A resource link for teachers and upper grade students.

The Purse, by Kathy Caple. Houghton Mifflin Company: Boston, 1986. (Recommended Grades: K-2)

Katie keeps her money in a Band-Aid box until her older sister convinces her to buy a purse. Because she uses all her money for the purse, she has nothing left to put in it. She does earn more money and the way she spends it provides a novel twist to the ending.

Pyramid, by David Macaulay. Houghton Mifflin Company: Boston, 1977. (Recommended grades: 5-up)

Text and black-and-white illustrations follow the intricate process of how the great pharaohs' burial places were conceived and constructed. A fabulous book for learning about the pyramids and their mysteries. The text is very reader friendly and illustrations provide visual reinforcement.

Simon and His Boxes, by Gilles Tibo. Tundra Books: Plattsburgh, New York, 1992. (Recommended grades: K-3)

Simon is a little boy who has impossible dreams. He finds some cardboard boxes in the forest and he builds homes, apartments, and cities for the animals. But the animals refuse to join him in his cities, and Simon learns that animals prefer the homes they already have, and that boxes can have other, better uses.

Sitting In My Box, by Dee Lillegard. Penguin Putnam Books: New York, 1993. (Recommended grades: K-2)

With only his imagination for company, a little boy sits alone in a cardboard box, and is soon joined by animal after animal, with hilarious results.

A Special Kind of Love, by Stephen Michael King. Scholastic, Inc.: New York, 1996. (Recommended grades: PreK-3)
Explores the relationship between a father and a son and the unique way in which the father expresses his love. He builds wonderful things for his son—an enormous castle, a colorful kite, and a speedy go-cart, all out of boxes.

Spiders Spin Webs, by Yvonne Winer. Charlesbridge Publishing: Watertown,MA, 1998. (Recommended grades: K-3)
Young readers get a chance to look up-close at a stunning variety of webs from around the world. Rhyming text describes how, when, where, and why spiders spin webs.

Super Structures, by Philip Wilkinson. Houghton Mifflin: Boston, 1996. (Recommended grades: 4-6)
An "Inside Guides" book on amazing feats of engineering and construction. It uses three-dimensional models to explore fascinating structures, revealing their inner workings. Learn what keeps a soaring skyscraper standing, and what record-breaking bridge is nearly twice the size of its nearest rival. Great resource for student project reports and teacher information.

Super Structures of the World, by Stewart Kallen. Abdo & Daughters, 1991. (Recommended grades: 4 and up)
Surveys notable structures built by humans from ancient dwellings to modern amusement parks.

The Statue of Liberty, by Lucille Recht Penner. Random House: New York, 1995. (Recommended grades: 1-3)
This "Step Into Reading – Step 1" book explains the significance of the Statue of Liberty, where it originated, how it was constructed, and how it came to be in America. It describes the size of the structure and the process for building the high pedestal.

The True Story of the Three Little Pigs, by A. Wolf, by Lane Smith, As Told to Jon Scieszka. Scholastic, Inc.: New York, 1989. (Recommended grades: PreK-3)
A retelling of the folktale, by the wolf, who then gives his own outlandish version of what really happened when he tangled with the three little pigs. Being a victim for centuries of bad press, Alexander T. Wolf (you can call him Al), comes forward to give his side of the story. He was framed! It's not his fault if two of the pigs built shoddy houses.

Unbuilding, by David Macaulay. Houghton Mifflin Company: Boston, 1986. (Recommended grades: 1-6)
It is not a work of nonfiction, but a work of fantasy; and not the story of the making of a skyscraper structure, but the story of the dismantling of one—the Empire State Building.

The Very Busy Spider, by Eric Carle. Putnam Publishing: New York, 1995. (Recommended grades: PreK-2)
Pictures to feel as well as to see while you hear or read. Children use their fingers to trace the growth of the spider's structure, the spider's web. The collages are striking, the repetitive phrases and imitative sounds appealing, and the tactile experience of the growing web is educationally rewarding.

Whatever Happened to Humpty Dumpty?: And Other Surprising Sequels to Mother Goose Rhymes, by David T. Greenberg. Little, Brown & Company: Boston, 1999. (Recommended grades: 1-6)
Outrageous gross-out sequels to 20 Mother Goose rhymes present packaging from a humorous point of view. What ever happened to Humpty Dumpty after he fell off that wall? How did Peter the Pumpkineater's wife get back at him for putting her in a pumpkin shell? Teachers will have fun using this book in poetry units.

What's In A Box?, by Kelly Boivin. Children's Press: Chicago, 1991. (Recommended grades: K-2)

A book in the beginning-to-read series, "Rookie Readers," is a good introduction to packaging for emergent readers. It describes different types of boxes, and what they may hold, using both verse and graphics.

When This Box Is Full, by Patricia Lillie. Greenwillow Books: New York, 1993. (Recommended grades: K-3)

Using an empty wooden box and twelve familiar items representing the months of the year, young children engage in inquiry-based activities. These concepts include: empty and full, inside and outside, collecting, counting, sharing, remembering and the passage of time.

Yard Sale, by James Stevenson. Greenwillow Books: New York, 1996. (Recommended grades: PreK-3)

The signs are up, the tags are on, and the animals of "Mud Flat" participate in a yard sale. In ten brief chapters, the many characters sell one another all sorts of packaged items. Very funny, enjoyable reading!

Assessment

Nearly everyone agrees about the importance of assessment, but what exactly is it, and why is it so significant in education? In a very broad sense, education is like a very large design problem and assessment is the method of evaluating the design. However, education has many objectives, not just one, so assessment also includes a complex process of deciding what to assess and how. Another major complication is that many different kinds of people have a stake in the outcome of the educational process. Parents want to know how much their children are learning and how they can best help them. Politicians worry about the backlash from voters if the educational system appears to be "failing," however that term is defined. Administrators fear that they will be held accountable for low test scores in their schools.

Teachers, who have the most sustained and direct involvement of any adults in the educational process, are constantly looking for ways of knowing how well and how much their students are learning. This data can come from both formal and informal assessment methods, and may be either qualitative or quantitative. At the same time, teachers are often held accountable to conflicting requirements that are difficult or impossible to meet. For example, the goal of providing a supportive and welcoming learning environment may be in conflict with the regimentation imposed administrative requirements. Another common

concern of teachers is that high-stakes testing will require them to "teach to the test" rather than to support student learning.

Regardless of demands from outside the classroom, a teacher's primary responsibility is to engage students in exploring and understanding the subject matter. Assessment includes any method of finding out how much of this exploring and understanding actually happens. Information gained through assessment is the only factual basis for knowing what students are learning, how to motivate learning more effectively, how and whether to redesign the curriculum, how to tailor it to the needs of individual students, and how and when to involve parents in the process. Assessment is far too extensive and important to be narrowly defined by standardized test results or to be determined by people outside the classroom.

Here are some basic conclusions that follow from this view of assessment:

- Assessment should be based on clear educational goals.
- Many different kinds of information should be collected as part of assessment. Some of the most important assessment data is totally unexpected.
- Assessment should not be divorced from curriculum; every learning activity should also provide information for assessment.

- Whenever possible, students should become involved in assessing their own learning—for example, by evaluating their own designs or predictions.
- Assessment should examine not only what students have learned, but also the opportunities provided by the curriculum and the learning environment.

The following account by Sandra Skea, a fifth grade teacher at Mott Hall School, shows how assessment opportunities can arise during the course of a curriculum unit. Sandra's story of the Portable Storage System appears in Chapter 4 (pp. 119-125).

The construction of a portable classroom storage system was a project we developed to solve a real design problem. We were in need of a storage system to hold shoebox dioramas. Because the class met in several different classrooms during the week, we needed a system that would be portable and manageable.

At the time, we were exploring the properties of rectangular prisms in math class. Some students were having difficulty calculating girth, volume, and surface area from two-dimensional drawings found in the text. Even those who could do the calculations were relying on formulas they might not understand and would not be likely to remember. So aside from giving my students an opportunity to plan and work together as they designed and tested solutions for our storage problem, my goals included engaging students in using math and in communicating mathematically. I hoped that their desire to solve a design problem would facilitate their desire to understand and own the math involved. I hoped that they would be able not only to read and interpret two-dimensional drawings of three-dimensional objects, but also to understand the math embedded in the formulas.

Initially I had planned to use classroom observations, class discussions, and the final design projects as bases for assessment. I planned to look at and assess how well the students worked together, planned their time, met the design criteria, and used the necessary math. As the project progressed, I discovered many more opportunities for assessment and I also saw many more ways to engage the students in learning and understanding the design process and math.

I used my observations of the students and their reflections at the end of each day to guide me as I created new assessments. For example, I designed homework assignments based on the questions they were asking each other in class:

- How can we better use our time?
- How can we make the box stronger?
- Will another kind of adhesive or tape work better?
- How do we know ten dioramas will fit in the storage units?

One day I overheard a student saying, "I could use a system like this to store my toys. My mother says my room is a mess!" I then devised a homework assignment asking the students to report how they could create and use a similar design for use at home.

At another point, I heard a student ask, "Who wants to be the math expert and what is the math expert supposed to do anyway?" I immediately made up a homework assignment asking the students to describe the role of each group member and to explain how the roles were assigned.

Noting how engaged the students became in responding to these homework assignments and in sharing their new ideas led me to explore other assessment methods. Since the students were so clearly proud of their work, I felt there should be a formal way for them to record their accomplishments. I decided to have each student write a final report. Students became very excited about this idea, because it would give them a chance to express the issues and ideas they had discovered. Many students included charts, graphs, and drawings. The students found these reports helpful in planning their group oral presentations.

As I circulated during work time and listened to the daily progress reports, I asked reflective questions. I learned where student understanding of various mathematical concepts was weak or incorrect. Asking students to look at parallel lines, perpendicular lines, angles, faces, and areas created learning opportunities for them and new teaching and assessment opportunities for myself.

For example, looking at a shelf in terms of angles and lines led students to discover that two right angles can combine to make a straight angle, which has 180 degrees. I discovered that a simple design problem presents countless opportunities for teaching geometric concepts, measurement, and number sense.

I introduced peer assessment and self-assessment after the design process was complete. The students evaluated the final projects, as each group presented their storage unit, findings, and ideas to the rest of the class. The student self-assessment questionnaire asked students to reflect on their storage system design, their participation in the design process, and the math and design principles they had learned.

Affording students the opportunity to engage in design and testing allowed me to assess their progress in becoming good problem-solvers. I was able to gain insight into how well they can identify a problem, make a plan, design, redesign, test and evaluate their solutions. These kinds of information are simply not available from a standardized test. A project-based assignment allows students to correct mistakes, and find out the math and design principles they need to solve a real problem. Students become good problem-solvers when they when they are allowed to explore their own solution methods.

The students' progress became apparent when they were later asked to explore the properties of a cylinder. A student suggested that the way to find the volume of a cylinder might be like the formula for finding the volume of a rectangular prism, such as a storage unit:

"If the base of the storage system times its height equals the volume of the box, maybe if we multiply the base of a cylinder times its height we will get its volume." It was clear that this student was able to connect a formula with a concept, and apply both in a new situation.

I learned from this project that we as teachers must be alert for new and alternative assessment possibilities as they arise. My goals for the project became much richer as I saw my students participate in a mathematical community that fostered exploring, interacting, and using math in a real-life context. The educational goals that evolved led to the greatest insight into how my students learn, and how they can use and communicate what they have learned. The key was that I began with a good problem, and that I allowed room for new goals and assessment methods, driven by the students' experiences in becoming good problem-solvers and masters of their own learning.

Educational Goals

In order to assess the learning outcomes of an activity, it is necessary to know what the educational goals were. However, the purpose of a curriculum unit may not be so clear-cut. Any worthwhile educational activity probably has more than one goal. Also, a teacher's goals may (and often do) change as the activity progresses, or there may be unintended outcomes that are far more significant than the original goals.

Sandra's account illustrates how the goals of an extended unit evolve as the unit progresses. Initially, Sandra saw the Portable Storage Units as a means of providing a real-world context for learning some math concepts that were difficult for her students to grasp. As the groups began to work on their designs, Sandra saw many more possible goals.

Students were learning ideas about structures, time management, group work, design and redesign that were at least as important as the original math goals. Consequently, she developed assessment methods to gauge their learning in these other areas as well. At the same time, Sandra felt that her math goals had been achieved. Evidence for this claim came after the project was over, when a student applied what she had learned from the project to a new problem: finding the volume of cylinder.

Rigid adherence to an initial set of goals assumes that the educational process is entirely predictable, which is not the case. Every teacher has both short- and long-term goals for her students, and it is difficult to know in advance when something will happen to advance the long-term goals unexpectedly. As Sandra put it during a discussion on assessment, "You can talk about goals all you want, but what I really care about is that they feel good about themselves and about what they are able to accomplish."

Information from a Variety of Sources

If educational goals are complex and multifaceted, so are the means of assessing to what extent these goals are met. The narrowest view of assessment, most popular in political circles, confines it to standardized tests. A somewhat broader view expands assessment to include all kinds of paper-and-pencil instruments designed specifically for assessment, such as worksheets, homework assignments, tests, and quizzes.

Our view of assessment is broader still. Let's look at the broad variety of assessment methods Sandra employed in the Portable Storage Structures Unit. Each day, at the end of the work session, each group was required to give a short account of their work during the day, and of their plans for the next day. Many of the issues raised during these sessions became the basis for homework assignments for the entire class. These assignments included drawing and reflective writing as well as performing calculations. The project culminated in a written final report by each student, and an oral presentation by each group. Each student also had to write a self-evaluation of his or her contributions to the project.

Part of the attraction of teaching is that much of what happens in the classroom is unpredictable, and some of the surprises are pleasant and even thrilling. Consequently, it is impossible to decide in advance what all of the methods of assessment will be. Often, serendipity provides ways of assessing students' learning that nobody could have anticipated.

Two striking examples of serendipity appear in Theresa Luongo's account of her pre-K/K class in Chapter 4 (pp. 92-99). Children who had been exploring pump dispensers decided to use them to "empty the smelly water from the water table." In other words, they saw the potential of the device to solve a new problem, one it had not been designed for. Even more significant, students who had been testing shopping bags decided to repair the broken bags. This decision led to other repair activities, such as repair of torn book covers. Clearly, these students had learned about more than the strength of bags. They became aware of their own potential as problem solvers and redesigners.

Curriculum as a Source of Assessment Data

In order to maximize the amount of information available, the curriculum itself must be seen as a rich source of assessment data. Verona Williams' story includes an account of how her students struggled with classifying bags

of different types. Samples of their work are shown in Figures 4-14 through 4-19, accompanied by Verona's commentary on each student's work (pp. 102-104). There is wide variation in what they did. Some students used more than one set of categories, but did not show how they are related, such as big and little plus paper and plastic. Another student mentioned a variety of categories, and also drew large and small bags, but did not count the bags in other categories. Another gave clear descriptions of the uses of three different kinds of bags, but did not sort bags or classify them. The one student who drew a bar graph of different kinds of bags did not use real data. The one accurate bar graph was of dolls, not bags. Each of these worksheets provides Verona with a window into a student's reasoning process, while also serving as a valuable learning experience.

Students Assess Their Own Learning

Should the audience for assessment data include students themselves? Obviously, students need to know how well they are doing, so they can gauge their own efforts and develop realistic goals for their own learning. However, traditional assessment is usually presented to students in an adversarial manner, in the form of test grades and report cards that frequently undermine rather than enhance their

motivation for learning. In traditional forms of assessment, students are always evaluated by adults rather than by themselves, and the outcomes of assessment often have high stakes. Both of these factors contribute to the view of assessment as an antagonistic process. How can students gain access to candid data about their own learning?

Some examples of peer- and self-assessment appear in Chapter 4. Christine Smith asked each student to fill out an evaluation sheet, which included both a self-evaluation and also a rating of each group member. Sandra Skea asked students to evaluate one another's oral group presentations, as well as daily reflections on the group process and a final self-evaluation form at the end.

In Sandra's and Michael Gatton's classes, there was an element of self-assessment that is rarely found in school settings. Both classes undertook projects intended to solve problems that were real to the students: the need for storage space. Both teachers remarked that the students had designed and constructed structures that were of real use to them, and that had actually solved their problems. The students could evaluate their own work by seeing these devices in daily use.

Assessing Teaching, Curriculum, and Environment

Like anybody else who designs or plans anything, most teachers engage in informal assessment of their work on an ongoing basis. They ask themselves, "Is it working?" This question is really one of self-assessment: "What is the quality of the learning opportunities I have provided for my students?" Some of this self-assessment by teachers is based on student learning outcomes of the many kinds described above. At the same time, teachers also assess learning opportunities on the basis of their own perceptions and experiences. Chapter 4 has several good examples of self-assessment and redesign of learning environments by teachers. At the end of her unit on packaging, Verona Williams saw several ways to improve it:

- Integrate sorting and classifying bags with a math unit on making charts and graphs;
- Replace some of the invididual activities with group activities;
- Provide time for examining how and where the bags broke during bag testing; and
- Include an activity on redesign of the bags to make them stronger.

Similarly, both Minerva Rivera and Christine Smith redesigned their packaging units in midstream to deal with classroom management issues. Minerva divided her class in two, and found another activity for those who had not become engaged in testing paper bags; while Christine devised an activity to engage those students who had already conducted their tests of pump and spray dispensers.

Sandra Skea has described at some length how she revised and extended Portable Storage Units, originally developed as a math activity, into something considerably more ambitious. Michael Gatton, by contrast, began with a very ambitious project, which he found he had to scale down to make it manageable. His concluding reflections include suggestions for trimming the project even further.

The Institutional Context

Every school is different. Each one offers both resources that can be helpful in implementing a new curriculum, and barriers that can make it difficult. It is useful to analyze both carefully, with an eye to mobilizing and extending the resources and overcoming the barriers. In this section, we will look at how some teachers have gained crucial support from school staff, parents, other teachers, and administrators as they developed new programs in science and technology.

The Custodian

The custodian is a key person in the success of any new program, which may take students outside of the classroom and into the rest of the building. The custodian is probably more familiar with the physical layout of the building than anyone else. He or she also has the best access to discarded materials, such as cardboard, waste paper, or wood, that can be very useful. A cooperative custodian can also offer suggestions about additional storage space, and can insure that projects in process will not be thrown out.

The custodian's involvement can also lead to exciting surprises, as the following story illustrates. A second-grade teacher and her class were studying the water supply system of a school in the South Bronx, New York City. They began with the water fountain just outside their classroom. The children were convinced that the water for the fountain was stored in the wall just behind it. Then somebody noticed that there were pipes leading to the fountain. They followed the pipes along the ceiling and realized that they came from someplace else in the building. At this point they went to another floor and noticed a similar pattern of pipes. Eventually, their investigation led them to the basement. There they met the custodian, who gave them copies of the blueprints (maps) of the building, and showed them how the water came into the building. The following day, he gave them an opportunity to turn on the boiler, so they could see how the hot water was heated! The outcome of this investigation was a working 3D model of the building's water supply, in which the pipes were represented by straws and the reservoir by a basin held above the highest floor.

The Parents

Parents can also be critical to the success of a curriculum project. A number of teachers have involved parents in investigations of the community around the school. One ESL teacher in East Harlem, New York City, whose students were recent immigrants from various parts of Latin America, engaged her students in a study of the casitas in the community. A casita (literally, "little house") is a small building constructed by community residents on a vacant lot, which may serve as a club house or a religious shrine, or which may be used to house livestock. Several parents who were very familiar with the community accompanied the class on

their field visits and facilitated their discussions with the users of the casitas.

How does a teacher get parents involved in the first place? Some teachers have organized parent/child workshops, after school or on Saturdays, as a way to inform parents of what their children are doing and to solicit their support. One strategy that has worked is to have a parent/child workshop a few weeks after children have begun a project. In the workshop, parents and their children are encouraged to pursue a hands-on project that is similar to what the children have already been doing in school. Because the children have already started the project, they will often take the lead in explaining the material and offer their parents advice on how to proceed. At the same time, parents will provide their own experiences and expertise, and some may become excited enough to volunteer additional support. Parent volunteers can provide the additional adult presence needed for taking the class outside the building.

Other Teachers

Just as children often require peer interaction to pursue a project, so peer support can be essential for teachers too. Another teacher can be a springboard for ideas, a source of advice on overcoming difficulties, and a friend to turn to when everything seems to go wrong. There are many models for teacher/teacher collaboration, each of which can work in some circumstances. Ultimately, the collabo-

rators have to figure out for themselves what works best for them. Here are some examples of ways in which teachers in the same school have worked together:

- An experienced teacher gave workshops in the school, in which she engaged other teachers in some of the same activities she had been doing in her classroom. Several of the other teachers became interested and sought advice on pursuing these activities in their own classrooms.
- An experienced special education teacher mentored a less experienced special ed teacher, offering her assistance in some of the same projects she had done in her own classroom.
- A science cluster teacher met with a classroom group during a "prep" period twice a week. She enlisted the students' classroom teacher in pursuing some of the same projects as part of their regular classroom work.
- A fifth grade teacher and a kindergarten teacher decided to work together. After the fifth-graders had pursued some of their own investigations, several of them became the facilitators in helping the kindergarten children do similar studies. Besides being a collaboration among teachers, this project was also a collaboration between older and younger children.

Collaboration among teachers may be actively discouraged by the culture of the school. Even in the best circumstances, collaborations can be difficult to sustain. Just as every school is different, so is every classroom. Ideas and strategies that work in one classroom may or may not be directly transferable to another, and it is important to remain sensitive to differences in chemistry and culture from one room to the next. The most important ingredient in a collaboration among teachers is the commitment to work and learn together, regardless of the outcome of any particular project or idea.

School Adminstration

A major component of a teacher's setting is the culture of the school administration. A principal, assistant principal, or other supervisor can make or break an innovative curriculum project. Some teachers are fortunate enough to find themselves in environments that nurture innovation; others are not so lucky. For better or worse, the tone set by the administration is a major factor that every teacher has to deal with. Even without initial support, however, there are a number of strategies for bringing a skeptical (or even a hostile) administrator on board. Here are some methods that have worked.

One teacher, who was a participant in an in-service inquiry science program, had a roomful of upper-elementary students engaged in long-term science investigations, largely of their own design. She decided to encourage them to enter their projects in the school science fair. She immediately ran into the opposition of her principal, who insisted that all of the material on the display boards be "professionally done." The teacher knew that her students were invested in their projects, and perfectly capable of creating their own displays, but unable to type the material or produce fancy graphics. To make the displays for them would be to undermine all of their efforts and enthusiasm. So she presented the situation to her children, without any suggestion about what they ought to do about it.

The next time the principal visited their classroom, the students let him know that they wanted to enter the science fair, and they believed they could make display boards which would be perfectly readable. In any case, they would be around to explain anything the judges didn't understand. With the teacher standing by silently, the principal reluctantly gave in. At the fair, it became clear that these were the students who had the best grasp of their own projects, although there were others that had nicer-looking boards. Neither the children nor the teacher were surprised when they won first, second and third prizes, and went on to the District fair! Equally important, the teacher felt that this was a turning point in her relationship with the principal. Afterwards, he interfered much less with her efforts at innovation.

It is far more effective to mobilize children, parents, other teachers, and staff than to confront an administrator directly. He or she will have a much harder time saying no to children, parents, or a group of teachers, than to a individual. Also, successful programs speak for themselves. Outside authorities, such as science fair judges, funding sources or important visitors, can make even the most reluctant principal sit up and take notice. Most important, innovation succeeds best when innovators lay the seeds quietly over time, and exploit opportunities to overcome resistance.

Resist the temptation to take on every adversary, every time. Focus instead on the resources that are available to you, and learn how to mobilize them effectively. Wait for opportunities to let your efforts speak for themselves.

Chapter 6

ABOUT STANDARDS

Overview

In Chapter 3, "Activities," we have listed standards references for each activity. This type of listing is now found in most curriculum materials, in order to demonstrate that the activities "meet standards." In a way, these standards references miss the point, because the national standards are not meant to be read in this way. Meeting standards is not about checking off items from a list. Each of the major standards documents is a coherent, comprehensive call for systematic change in education.

This chapter shows how *Stuff That Works!* is consistent with national standards at a very fundamental level. We will look in detail at the following documents:

- *Standards for Technological Literacy: Content for the Study of Technology* (International Technology Education Association, 2000);

- *Benchmarks for Science Literacy* (American Association for the Advancement of Science, 1993);

- *National Science Education Standards* (National Research Council, 1996);

- *Principles and Standards for School Mathematics* (National Council of Teachers of Mathematics, 2000); and

- *Standards for the English Language Arts* (National Council of Teachers of English & International Reading Association, 1996).

Most of these standards are now widely accepted as the basis for state and local curriculum frameworks. The first document on the list is included because it is the only national standard focused primarily on technology. The *New Standards Performance Standards* (National Center on Education and the Economy, 1997) is not included because it is based almost entirely on the *Benchmarks, National Science Education Standards*, the original NCTM *Math Standards* (1989), and the *Standards for the English Language Arts*.

Although they deal with very different disciplines, these major national standards documents have many remarkable similarities:

- They are aimed at *all* students, not only those who are college-bound.

- Using terms like "literacy" and "informed citizen," they argue that education should prepare students to understand current issues and participate in contemporary society.

- They recommend that school knowledge be developed for its use in solving real problems rather than as material "needed" for passing a test. They strongly endorse curriculum projects that arise from students' own ideas, experiences, and interests.

- They focus on the "big ideas" of their disciplines as opposed to memorization of isolated facts or training in narrowly defined skills. In other words, fewer concepts should be dealt with in greater depth. As the *National Science Education Standards* express it, "Coverage of great amounts of trivial, unconnected information must be eliminated from the curriculum." (NRC, 1996, p. 213)

• The standards reject standardized tests as the sole or even the major form of assessment. Traditional exams measure only what is easy to measure rather than what is most important. "While many teachers wish to gauge their students' learning using performance-based assessment, they find that preparing students for machine-scored tests – which often focus on isolated skills rather than contextualized learning – diverts valuable classroom time away from actual performance." (NCTE/IRA, 1996, p. 7) The standards promote authentic assessment measures, which require students to apply knowledge and reasoning "to situations similar to those they will encounter outside the classroom." (NRC, 1996, p. 78) Furthermore,

assessment should become "a routine part of the ongoing classroom activity rather than an interruption" and it should consist of "a convergence of evidence from different sources." (NCTM, 2000, p. 23)

• They highlight the roles of quantitative thinking, as well as oral and written communication, in learning any subject, and they emphasize the inter-disciplinary character of knowledge.

• They view learning as an active process requiring student engagement with the material and subject to frequent reflection and evaluation by both teacher and learner.

• They urge teachers to "display and demand respect for the diverse ideas, skills and experiences of all students,"

and to "enable students to have a significant voice in decisions about the content and context of their work." (NRC, 1996, p. 46)

The *Stuff That Works!* materials are based on these ideas and provide extensive guidance on how to implement them in the classroom. We begin our study of technology with students' own ideas and experiences, address problems that are of importance to them, develop "big ideas" through active engagement in analysis and design, and draw connections among the disciplines. While the standards are clear about the principles, they do not provide many practical classroom examples. *Stuff That Works!* fills this gap.

Where the Standards Came From

Historically speaking, the current tilt towards national curriculum standards is a dramatic departure from a long tradition of local control of education. How did national standards manage to become the order of the day? In the late 1970's, the country was in a serious recession, driven partly by economic competition from Europe and Japan. In 1983, the National Commission on Educational Excellence (NCEE) published an influential report,

A Nation at Risk, which painted a depressing picture of low achievement among the country's students. The report warned of further economic consequences should these problems continue being ignored, and advocated national curriculum standards for all students. Adding to these arguments were pressures from textbook publishers, who felt that national standards would make state and local adoption processes more predictable.

Around the same time, several of the major professional organizations decided to provide leadership in setting standards. The pioneering organiza-tions were the National Council of Teachers of Mathematics (NCTM) and the American Association for the Advancement of Science (AAAS), whose efforts culminated in the publication of major documents in 1989. In the same year, the National Governors' Association and the first

Bush Administration both endorsed the concept of establishing national educational goals. The NCTM was deeply concerned about the issues raised by *A Nation at Risk* and was convinced that professional educators needed to take the initiative in setting a new educational agenda. Otherwise, the reform of curriculum would rest in the hands of textbook and test publishers, legislatures, and local districts.

Both the NCTM and the AAAS standards projects began with similar basic positions about pedagogy. Influenced by research about what children actually know, they recognized the disturbing fact that "learning is not necessarily an outcome of teaching." (AAAS, 1989, p. 145) In contrast with traditional approaches to education, which emphasize memorization and drill, the new national standards promote

strategies for active learning. A related theme of the early standards efforts was that the schools should teach fewer topics in order that "students end up with richer insights and deeper understandings than they could hope to gain from a superficial exposure to more topics…" (p. 20)

Meeting standards requires a major investment of time and resources. Some of the necessary ingredients include new curriculum ideas and materials, professional development opportunities, new assessment methods, and smaller class sizes. The *National Science Education Standards* are the most explicit in identifying the conditions necessary—at the classroom, school, district, and larger political levels—for standards to be meaningful. The authors state, "Students could not achieve standards in most of today's schools." (NRC,

1996, p. 13) More money might not even be the hardest part. Standards-based reforms also require understanding and commitment from everyone connected with the educational system, starting at the top.

The history of standards may contain clues about their future. Standards imply neither textbook-based instruction nor standardized tests. Standards arose because traditional text- and test-based education had failed to result in the learning of basic concepts by the vast majority of students. Ironically, there are many textbook and test purveyors who market their products under the slogan "standards-based." Standards could easily become discredited if those who claim their imprimatur ignore their basic message.

What the Standards Actually Mean

Standards are commonly read as lists of goals to be achieved through an activity or a curriculum. This approach is reflected in the lists of standards references and cross-references that appear in most curriculum materials, as evidence that an activity or curriculum "meets standards."

Presenting lists of outcomes in this fashion reflects a narrow reading

of standards, which can be very misleading. These lists suggest that "meeting standards" is simply a matter of getting students to repeat something like the statements found in the standards documents.

In fact, the standards are much richer and more complex than these lists imply. Many of the standards do not even specify the knowledge that

students should acquire, but deal rather with ways of using that knowledge. Here is an example from *Benchmarks for Science Literacy*:

"By the end of fifth grade, students should be able to write instructions that students can follow in carrying out a procedure." (p. 296)

This standard talks about something students should be *able to do,* rather than what they should *know.* The newly released NCTM document, *Principles and Standards for School Mathematics* (2000), unlike the earlier one (NCTM, 1989), explicitly separates "Content Standards" from "Process Standards." The Content Standards outline what students should learn, while the Process Standards cite ways of acquiring and expressing the content knowledge. The Process Standards include problem solving, communication, and representation. The *Benchmarks* example cited above is another example of a process standard. Similarly, in the English Language Arts (ELA) document (NCTE/IRA, 1996), all twelve standards use verbs to express what students should *do,* as opposed to what they should *know.* Any reading of standards that focuses only on content knowledge is missing a central theme of all of the major documents.

There is also material in the standards that qualifies neither as content nor as process. Here is an example from the *Benchmarks* chapter called "Values and Attitudes":

"By the end of fifth grade, students should raise questions about the world around them and be willing to seek answers to some of them by making careful observations and trying things out." (p. 285)

This standard asks for more than a specific piece of knowledge, ability, or skill. It calls for a way of looking at the world, a general conceptual framework, that transcends the boundaries of disciplines. Similarly, the "Connections" standard in the new NCTM document underscores the need for students to ...

"...recognize and apply mathematics in contexts outside of mathematics." (NCTM, 2000, p. 65)

These are examples of broad curriculum principles that cut across the more specific content and process standards. These standards are not met by implementing a particular activity or by teaching one or another lesson.

They require an imaginative search for opportunities based on a reshaping of goals for the entire curriculum.

In general, the standards documents are at least as much about general principles as about particular skills and knowledge bases. The *Standards for Technological Literacy,* the *Benchmarks,* and the *National Science Education Standards* each identifies some big ideas that recur frequently and provide explanatory power throughout science and technology. "Systems" and "modeling" are concepts that appear in all three documents. The presence of such unifying ideas suggests that the individual standards references should not be isolated from one another. They should rather be seen as parts of a whole, reflecting a few basic common themes.

What Use Are Standards?

Increasingly, teachers are being held accountable for "teaching to standards." These demands are added to such other burdens as paperwork, test schedules, classroom interruptions, inadequate space and budgets, arbitrary changes in class roster, etc. In the view of many teachers, children and their education are routinely placed dead last on the priority list of school officials. Understandably, teachers may resent or even resist calls to "meet standards" or demonstrate that their curricula are "standards-bearing." It is not surprising that many teachers cynically view the standards movement as "another new thing that will eventually blow over."

The push to "meet standards" is often based on a misreading of standards as lists of topics to be "covered" or new tests to be administered. It is not hard to imagine where this misinterpretation might lead. If the proof of standards is that students will pass tests, and students fail them nevertheless, then the standards themselves may eventually be discarded. Paradoxically, the prediction that "this, too, shall pass" would then come true, not because the standards failed, but because they were never understood nor followed.

Standards are intended to demolish timeworn practices in education. Some of these practices place the teacher at the center of the classroom but reduce

her or him to a cog in the machinery of the school and the district, with the primary responsibility of preparing students for tests. The standards documents recognize the need to regard teachers as professionals, students as active, independent learners, and tests as inadequate methods of assessing the full range of learning.

Within broad frameworks, the standards urge teachers to use their judgment in tailoring the curriculum to students' needs and interests. The NRC *Science Standards*, for example, call for "teachers [to be] empowered to make the decisions essential for effective learning." (1996, p. 2) Neither teachers nor administrators should interpret standards as mechanisms for tightening control over what teachers and students do. While they are very clear about the goals of education, the standards are less specific about how to meet them. Innovative curriculum efforts such as *Stuff That Works!* fit very well within the overall scheme of standards.

Teachers who have tried to implement *Stuff That Works!* activities in their classrooms have often come away with a positive feeling about them. The following comments are typical:

• *The strengths of this unit are the opportunity to group students, work on*

communication skills, problem solve … and plan real life tests. I have watched my students go from asking simple yes/no questions to thinking and planning careful, thoughtful active questions. The students began to see each other as people with answers… I was no longer the expert with all the answers.

• *I must begin by telling you that I found this particular guide to be so much fun and the students demonstrated so much energy and interest in this area… I was able to engage them in the activities easily… The activities were very educational and provided so much vital information that helped students connect what is being taught to them in math to real life situations, e.g., graphing behavior and using tallies to record information. For my [special education] students, I found this gave them self confidence…*

• *I read the entire guide from front to back… Although the main idea of the unit is not specifically a large focus of instruction in our fourth grade curriculum, I recognized the power behind the ideas and activities and knew that this unit would promote collaboration, problem solving and communication… Overall, I think my students loved this unit and felt enormously successful after we finished…*

• *My most important goal for students is that they feel good about themselves and realize what they can do. I liked these activities, because they had these results.*

The standards are intended to promote just these sorts of outcomes. When a teacher has a "gut feeling" that something is working well, there is usually some basis to this feeling. As the NRC Science Standards state, "outstanding things happen in science classrooms today… because extraordinary

teachers do what needs to be done *despite conventional practice* [emphasis added]." (1996, p. 12) Unfortunately, even an extraordinary teacher may not find support from traditional administrators, who complain that the classroom is too noisy or messy, or that somebody's guidelines are not being followed. Under these circumstances, standards can be very useful. It is usually easy to see how valuable innovations fit into a national framework of education

reform that is also endorsed by state- and district-level authorities. Standards can be used to justify and enhance innovative educational programs whose value is already self-evident to teachers and students.

What the Standards Really Say

In order to justify work as meeting standards, it is necessary to know what the standards really say. In the remainder of this chapter, we summarize each of the five major standards documents listed at the beginning of the chapter, and show how the *Stuff That Works!* ideas are consistent with these standards. We provide some historical background for each of the standards, and look at the overall intent and structure before relating them to the *Stuff That Works!* materials. These sections should be used only as they are needed. For example, if you would like to use some of the ideas from this Guide, and are also required to meet the *National Science Education Standards,* then that section could be useful to you in helping you justify your work.

Standards for Technological Literacy: Content for the Study of Technology

In April 2000, the International Technology Education Association (ITEA) unveiled the *Standards for Technological Literacy,* commonly known as the *Technology Content Standards,* after extensive reviews and revisions by the National Research Council (NRC) and the National Academy of Engineering (NAE). In its general outlines, the new standards are based on a previous position paper, *Technology for All Americans.* (ITEA, 1996) The latter document defined the

notion of "technological literacy" and promoted its development as the goal of technology education.

A technologically literate person is one who understands "what technology is, how it is created, and how it shapes society, and in turn is shaped by society." (ITEA, 2000, p. 9) According to the *Standards,* familiarity with these principles is important not only for those who would pursue technical careers, but for all other students as well. They will need to know about technology in order to be thoughtful practitioners in most fields, such as medicine, journalism, business, agriculture, and education. On a more general level, technological literacy is a requirement for participation in society as an intelligent consumer and an informed citizen.

Given the importance of being technologically literate, it is ironic that technology barely exists as a school subject in the U.S., and is particularly hard to find at the elementary level. In a curriculum overwhelmingly focused on standardized tests, there seems to be little room for a new subject such as technology. To make matters worse, there is considerable confusion over what the term *technology* even means. Many in education still equate it with "computers." The *Standards* advocate for technology education based on a broad definition of "technology," which is "how humans modify the world around them to meet their needs and wants, or to solve practical problems." (p. 22)

The *Technology Content Standards* describe three aspects of developing technological literacy: learning *about* technology, learning to *do* technology, and technology as a theme for curriculum integration (pp. 4-9). To learn about technology, students need to develop knowledge not only about specific technologies (Standards 14 – 20), but also about the nature of technology in general (Standards 1 – 3), including its core concepts: **systems, resources, requirements, trade-offs, processes,** and **controls**. Resources include **materials, information,** and **energy,** while **modeling** and **design** are fundamental examples of processes (p. 33). Students learn to "do" technology by engaging in a variety of technological processes, such as **troubleshooting, research, invention, problem solving, use and**

maintenance, assessment of technological impact, and, of course, **design** (Standards 8 – 13). Technology has obvious and natural connections with other areas of the curriculum, including not only math and science, but also language arts, social studies, and the visual arts.

Another set of standards deals with the relationship between technology and society. Some of the most basic understandings are that "products are made to meet individuals' needs and wants" (p. 74) and "some materials can be reused or recycled" (p. 66). Both of these ideas are addressed directly by the activities in *Packaging and Other Structures*. As students examine some packaging materials they have brought in, they begin by asking of each one: "What problems is this package designed to solve?" This question gets at the role of a very familiar form of technology in addressing human needs. A related question, which introduces reuse of technology, is: "What other purposes could this packaging material be used for?" This question often leads to design activities. For example, discarded packaging can become the material for creating useful classroom structures, such as storage space.

According to the Technology Content Standards, design is "the core problem-solving process [of technology]. It is as fundamental to technology as inquiry is to science and reading is to language arts." (p. 91) The importance of design is underlined by the

statement, a little further on, that "students in grades K-2 should learn that everyone can design solutions to a problem." (p. 93) Several pages later, the Standards suggest that young children's experiences in design should focus on "problems that relate to their individual lives, including their interactions with family and school environments." (p.100) However, the Technology Content Standards offer little if any guidance on how to identify such problems. The vignette provided on the following page, "Can you Help Mike Mulligan?", is based on a literature connection rather than children's environments.

Where does technology education "fit" in the existing curriculum? The Technology *Standards* address this problem by claiming that technology can enhance other disciplines: "Perhaps the most surprising message of the *Technology Content Standards* … is the role technological studies can play in students' learning of other subjects." (p. 6) We support this claim in the following sections, which draw the connections between *Stuff That Works!* and national standards in science, math, and English language arts.

Benchmarks for Science Literacy

There are two primary standards documents for science education: The American Association for the

Advancement of Science (AAAS) *Benchmarks for Science Literacy* (1993) and the National Research Council (NRC) *National Science Education Standards* (1996). Unlike the *National Science Education Standards,* the *Benchmarks* provide explicit guidance for math, technology, and social science education, as well as for science. The *Benchmarks* draw heavily on a previous AAAS report, *Science for All Americans* (1989), which is a statement of goals and general principles rather than a set of standards. *Benchmarks* recast the general principles of *Science for All Americans* (SFAA) as minimum performance objectives at each grade level.

The performance standards in *Benchmarks* are divided among 12 chapters. These include three generic chapters regarding the goals and methods of science, math and technology; six chapters providing major content objectives for the physical, life, and social sciences, technology, and mathematics; and three generic chapters dealing with the history of science, "common themes," and "habits of mind." The last four chapters of *Benchmarks* provide supporting material, such as a glossary of terms and references to relevant research.

Recognizing that standards are necessary but not sufficient for education reform, the AAAS has also developed some supplementary documents to support the process of curriculum change. These include *Resources for Science Literacy: Professional Development* (1997), which suggests reading materials for teachers, presents outlines of relevant teacher education courses, and provides comparisons between the *Benchmarks,* the Math Standards, the Science Standards and the Social Studies Standards. A subsequent publication, *Blueprints for Science Reform* (1998) offers guidance for changing the education infrastructure to support science, math, and technology education reform. The recommendations in *Blueprints* are directed towards administrators, policy makers, parent and community groups, researchers, teacher educators, and industry groups. A subsequent AAAS document, *Designs for Science Literacy* (2001), provides examples of curriculum initiatives that are based on standards.

The *Benchmarks* present a compelling argument for technology education. The authors present the current situation in stark terms: "In the United States, unlike in most developed countries in the world, technology as a subject has largely been ignored in the schools." (p. 41) Then they point out the importance of technology in children's lives, its omission from the curriculum notwithstanding:

"Young children are veteran technology users by the time they enter school.... [They] are also natural explorers and inventors, and they like to make things." (p. 44) To resolve this contradiction, "School should give students many opportunities to examine the properties of materials, to use tools, and to design and build things." (p. 44)

Like the Technology Standards, the *Benchmarks* identify **design** as a key process of technology and advocate strongly for first-hand experience in this area. "Perhaps the best way to become familiar with the nature of engineering and design is to do some." (p. 48) As children become engaged in design, they "begin to enjoy challenges that require them to clarify a problem, generate criteria for an acceptable solution, try one out, and then make adjustments or start over again with a newly proposed solution." (p. 49) These statements strongly support the basic approach of *Stuff That Works!,* which is to engage children in analysis and design activities based on the technologies already familiar to them. Like *Stuff That Works!,* the *Benchmarks* also recognize the back-and-forth nature of design processes, which rarely proceed in a linear, predictable sequence from beginning to end.

Work with packaging engages students in exploring the characteristics of materials: "Young children should have many experiences in working with different kinds of materials, identifying and composing their properties and figuring out their suitability for different purposes."(p. 188) Children develop these skills as they examine and test various packages and containers and figure out how to repair or strengthen boxes and bags. A further set of activities from *Packaging and Other Structures* involves them in thinking about how to reuse discarded packaging materials: "Students can reflect on the influences that their own consumption choices can have on what products are made and how they are packaged." (p. 189)

The analysis of containers, bags and packages, as described in *Packaging and Other Structures,* also develops the concepts of trade-offs and failure, which are both central to the "Nature of Technology." Benchmarks explain trade-offs as follows: "Designs that are best in one respect may be inferior in other ways. Usually some features are sacrificed in order to get others. How such trade-offs are received depends upon which features are emphasized and which are downplayed." (p. 49) For example, as children test a variety of shopping bags, it becomes obvious

that the one that is strongest when dry is probably not the best when soaking wet.

Closely related to trade-offs is the notion of failure. According to Benchmarks, "Even a good design may fail. Sometimes steps can be taken to reduce the likelihood of failure, but it can never be entirely eliminated." (p. 50) Testing shopping bags, a *Packaging and Other Structures* activity, is a natural and obvious way to develop an understanding of failure. Even in early childhood classes, children become aware that not all bags are the same, that they fail in different ways and that if loaded sufficiently, all of them will fail eventually. From the pre-K/K level upwards, testing shopping bags leads naturally to the repair and redesign of bags and other items. Through these activities, students "develop skill and confidence in using ordinary tools for personal and everyday purposes." (p. 45) Here are compelling reasons for engaging children with packaging: these activities will provide them with the experience and confidence to make, analyze, and fix things.

The National Science Education Standards

In 1991, the National Science Teachers Association asked the National Research Council to develop a set of national science education standards. These standards were intended to complement the *Benchmarks*, which include math, technology, and social studies as well as natural science. The National Research Council (NRC) includes the National Academy of Sciences, which is composed of the most highly regarded scientists in the country. Over the course of the next five years, the NRC involved thousands of scientists, educators, and engineers in an extensive process of creating and reviewing drafts of the new science standards. The results were published in 1996 as the *National Science Education Standards* (NSES).

Who is the audience for standards? The conventional view is that standards outline what students should know and be able to do, and that teachers are accountable for assuring that their students meet these guidelines. The NSES take a much broader approach, looking at the whole range of systemic changes needed to reform science education. The document is organized into six sets of standards. Only one of the six, the "Science Content Standards," talks directly about what children should learn through science education. The remaining five address other components of the education infrastructure, including classroom environments, teaching methods, assessment, professional development, administrative support at the school and district levels, and policy at the local, state, and national levels.

Collectively, these standards outline the roles of a large group of people on whom science education depends: teachers, teacher educators, staff developers, curriculum developers, designers of assessments, administrators, superintendents, school board members, politicians, informed citizens, and leaders of professional associations. If an administrator or school board member were to ask a teacher, "What are you doing to address the *National Science Education Standards?*" the teacher would be fully justified in responding, "What are *you* doing to meet them?"

One message that recurs frequently in the NSES is that teachers must be regarded as professionals, with a vital stake in the improvement of science education and an active role "in the ongoing planning and development of the school science program."

(p. 50) More specifically, they should "participate in decisions concerning the allocation of time and other resources to the science program." (p. 51) The *Standards* explicitly reject the reduction of teachers to technicians or functionaries who carry out somebody else's directives. "Teachers must be acknowledged and treated as professionals whose work requires understanding and ability." The organization of schools must change too: "School leaders must structure and sustain suitable support systems for the work that teachers do." (p. 223)

Teachers should also play a major role in deciding and/or designing the science curriculum. The *Standards* call for teachers to "select science content and adapt and design curricula to meet the needs, interests, abilities and experiences of students." Although teachers set the curriculum initially, they should remain flexible: "Teaching for understanding requires responsiveness to students, so activities and strategies are continuously adapted and refined to address topics arising from student inquiries and experiences, as well as school, community and national events." (p. 30) Not only teachers, but also students, should play a major role in curriculum planning. The Teaching Standards make

this point explicit: "Teachers [should] give students the opportunity to participate in setting goals, planning activities, assessing work and designing the environment." (p. 50)

More specifically, Content Standard E, "Science and Technology," strongly supports the approach of *Stuff That Works!:* "Children's abilities in technological problem solving can be developed by firsthand experience in tackling tasks with a technological purpose. They can also study technological products and systems in their world—zippers, coat hooks, and can openers… They can study existing products to determine function and try to identify problems solved, materials used and how well a product does what it is supposed to do… Tasks should be conducted within immediately familiar contexts of the home and school." (p. 135)

The Science Standards do not make the distinction between design and inquiry as sharply as do the Technology Standards: "Children in grades K-4 understand and can carry out design activities earlier than they can inquiry activities, but they cannot easily tell the difference between the two, nor is it important whether they can." (p. 135) Thus, many of the abilities and concepts needed to meet the standard "Science as Inquiry" are also developed through design. These include: "Ask a question about objects… in the environment"; "plan and conduct a simple investigation"; "employ simple equipment and tools to gather data"; and "communicate investigations or explanations." (p. 122)

The material in *Packaging and Other Structures* is of particular relevance to the K-12 Content Standards, "Unifying Concepts and Processes." One of the five unifying themes is "form and function." Students ask how the shape and size of a box, bottle, bag, or pump dispenser is related to the function it serves. Often, they may redesign a package to serve a somewhat different function; or they may design and create storage spaces from discarded packaging materials. Each of these activities provides numerous opportunities to learn that "the form or shape of an object or system is frequently related to use, operation or function." (p. 119) The ideas in this volume also address Content Standard B, "Physical Science," for grades K-4. Sorting and classifying packaging materials, for example, lead to multiple discoveries regarding the "properties of objects and materials" and the "similarities and differences of the objects." (p. 125)

A major upper-grade theme of *Packaging and Other Structures* is the design and construction of useful class-

room structures. These activities involve students in improving the use of space in their own classrooms. The "Teaching Standards" section of the NSES calls for just this sort of involvement:

"As part of challenging students to take responsibility for their learning, teachers [should] involve them in the design and management of the learning environment. Even the youngest students can and should participate in discussions and decisions about using time and space for work." (p. 45)

Principles and Standards for School Mathematics

The first of the major standards documents, *Curriculum and Evaluation Standards for School Mathematics*, was published in 1989 by the National Council of Teachers of Mathematics (NCTM). Additional standards for teaching and assessment were published in 1991 and 1995, respectively. In 2000, the NCTM released a new document, *Principles and Standards for School Mathematics*, intended to update and consolidate the classroom-related portions of the three previous documents. Some of the major features

of the new volume, different from the prior version, are the addition of the Principles, the division of the standards into the categories "Content" and "Process," and the inclusion of a new process standard called "Representation."

The new NCTM document acknowledges the limitations of educational standards: "Sometimes the changes made in the name of standards have been superficial or incomplete… Efforts to move in the direction of the original NCTM Standards are by no means fully developed or firmly in place." (pp. 5-6) In spite of this candid assessment, the authors remain optimistic about the future impact of standards. Their goal is to provide a common framework for curriculum developers and teachers nationwide. If all schools follow the same standards, then teachers will be able to assume that "students will reach certain levels of conceptual understanding and procedural fluency by certain points in the curriculum." (p. 7)

The NCTM *Principles and Standards* begin by presenting the six sets of principles, which are the underlying assumptions for the standards. Some of these principles are common to the other standards documents: that there should be high expectations of all students, that the goal of learning is deep understanding, and that assessment should be integrated with curriculum. Other principles underscore the need to learn from cognitive research. More than in any other field, there has been extensive research into how students learn mathematics, and this research base is reflected in the *Principles*. For example, the "Curriculum Principle" calls for coherent sets of lessons, focused collectively on one "big idea." Similarly, the "Teaching Principle" specifies that teachers must be aware of students' cognitive development. The "Learning Principle" cites research on how learning can be most effective.

The standards themselves are organized into two categories: Content Standards and Process Standards. The former describe what students should learn, in the areas of Number and Operations, Algebra, Geometry, Measurement, and Data Analysis and Probability. The Process Standards discuss how students should acquire and make use of the content knowledge. The subcategories are Problem Solving, Reasoning and Proof, Communication, Connections, and Representation. Unlike the earlier NCTM document, the new version uses all the same standards across all of the grade levels, from K through 12. In this way, the NCTM is advocating for a carefully structured curriculum, which builds upon and extends a few

fundamental ideas systematically across the grades. Readers may be surprised to find an Algebra Standard for grades K-2, or a Number and Operations Standard for grades 9-12.

Stuff That Works! units and activities offer rich opportunities for fulfilling a key ingredient of the NCTM standards: learning and using mathematics in

context. The Process Standard called "Connections" makes it clear that mathematics should be learned by using it to solve problems arising from "other subject areas and disciplines" as well as from students' daily lives." (p. 66) *Stuff That Works!* fulfills this standard in two fundamental respects: it provides mathematics connections with another subject area, technology, and it uses artifacts and issues from everyday life as the source of material for study. The mathematics students learn is drawn from all of the Content Standards, as well as all of the Process Standards except for Reasoning and Proof.

Sorting and classifying bags, boxes and bottles are very popular starting activities in *Packaging and Other Structures*. These activities prepare the way for the more formal methods of pattern handling known as algebra. The NCTM strongly recommends that these basic ideas about patterns be

developed with very young children. The Algebra Standard for grades K-2 calls for pattern finding and pattern recognizing activities, such as classifying and sorting, and identifying "the criteria [students] are using as they sort and group objects." Basic classifying activities are designed to "help students develop the ability to form generalizations." (p. 91) As part of *Packaging and Other Structures,* students sort mechanisms, switches, boxes, containers, or bags, and ask other students to "guess what our categories were" just by looking at the objects in each group.

More advanced activities engage students in other aspects of the NCTM Standards. For example, as students and sketch their designs for useful storage structures, or redesign a box to fit an object of a different size, they develop the basic techniques of "visualization, spatial reasoning and geometric modeling" (p. 43) that are central to the Geometry Standard. As they assess and refine their own solutions to design problems, they are meeting important aspects of the Problem Solving Standard. Collecting and presenting data about bag or pump dispenser tests, in graphic and verbal formats, are ways of addressing the Data Analysis and Probability Standard, which "recommends that students formulate questions that can be answered using data and addresses what is involved in gathering and using data wisely. Students should learn how to collect data, organize their own or others' data, and display the data in graphs or charts that will be useful in answering their questions." (p. 48)

Standards for the English Language Arts

By 1991, it had become clear that standards would be produced for all of the major school subjects. Fearful that English language standards might be produced without a firm basis in research and practice, two major professional organizations requested Federal funding for their own standards effort. The following year, the Department of Education awarded a grant for this purpose to the Center for the Study of Reading at the University of Illinois, which agreed to work closely with the two organizations, the National Council of Teachers of English (NCTE) and the International Reading Association (IRA). This effort culminated in the 1996 publication of the *Standards for the English Language Arts* by the NCTE and IRA. These ELA Standards are now widely accepted

for their clear, concise outline of English language education.

The ELA *Standards* adopt an unusually comprehensive view of "literacy," much broader in its scope than the traditional definition of "knowing how to read and write." (p. 4) Literacy also includes the ability to think critically, and encompasses oral and visual, as well as written communication. Recognizing that these forms of language "are often given limited attention in the curriculum," the *Standards* outline the variety of means used to convey messages in contemporary society:

> "Being literate in contemporary society means being active, critical, and creative users not only of print and spoken language, but also of the visual language of film and television, commercial and political advertising, photography, and more. Teaching students how to interpret and create visual texts such as illustrations, charts, graphs, electronic displays, photographs, film and video is another essential component of the English language arts curriculum." (pp. 5-6)

According to the ELA *Standards*, there are three major aspects to language learning: **content**, **purpose**, and **development**. Content standards address only *what* students should

learn, but not why or how: "Knowledge alone is of little value if one has no need to – or cannot – apply it." The *Standards* identify four purposes for learning and using language: "for obtaining and communicating information, for literary response and expression, for learning and reflection, and for problem solving and application." (p. 16) Purpose also figures prominently in the third dimension of language learning, development, which describes *how* students acquire this facility. "We learn language not simply for the sake of learning language; we learn it to make sense of the world around us and to communicate our understanding with others." (p. 19)

Of course, purpose and motivation vary from one situation to another. The authors of the *Standards* make this point, too, in their discussion of context: "Perhaps the most obvious way in which language is social is that it almost always relates to others, either directly or indirectly: we speak to others, listen to others, write to others, read what others have written, make visual representations to others and interpret their visual representations." Language development proceeds through the practice of these communication skills with others: "We become participants in an increasing number of language groups that necessarily influence the ways in which we speak, write and represent." While language development is primarily social, there is an individual dimension as well: "All of us draw on our own sets of experiences and strategies as we use language to construct meaning from what we read, write, hear, say, observe, and represent." (p. 22)

How does this broad conception of literacy and its development relate to daily classroom practice? The authors recognize that the ELA *Standards* may be in conflict with the day-to-day demands placed on teachers. "They may be told they should respond to the need for reforms and innovations while at the same time being discouraged from making their instructional practices look too different from those of the past." Among those traditional practices are the use of standardized tests, "which often focus on isolated skills rather than contextualized learning." Prescribed texts and rigid lesson plans are further barriers to reform, because they tend to preclude "using materials that take advantage of students' interests and needs" and replace "authentic, open-ended learning experiences." (p. 7) Another problem is "the widespread practice of dividing the class day into separate periods [which] precludes integration among the English language arts and other subject

areas." (p. 8) Taken seriously, these standards would lead to wholesale reorganization of most school experiences.

This introductory material sets the stage for the twelve content standards, which define "what students should know and be able to do in the English language arts." (p. 24) Although these are labeled "content" standards, "content cannot be separated from the purpose, development and context of language learning." (p. 24) In a variety of ways, the twelve standards emphasize the need to engage students in using language clearly, critically and creatively, as participants in "literacy communities." Within these communities, students sometimes participate as *receivers* of language—by interpreting graphics, reading and listening and—and sometimes as *creators*—by using visual language, writing, and speaking.

Some teachers have used the *Stuff That Works!* activities and units primarily to promote language literacy, rather than for their connections with math or science. Technology activities offer compelling reasons for children to communicate their ideas in written, spoken, and visual form. In early childhood and special education classrooms, teachers have used *Stuff That Works!* to help children overcome difficulties in reading and writing, because it provides natural and non-

threatening entry points for written expression. In the upper elementary grades, *Stuff That Works!* activities offer rich opportunities for students to want to use language for social purposes. Several characteristics of *Stuff That Works!* contribute to its enormous potential for language learning and use:

- Every unit begins with an extensive group discussion of what terms mean, how they apply to particular examples, how to categorize things, and/or what problems are most important.

- The activities focus on artifacts and problems that engage children's imaginations, making it easy to communicate about them. Teachers who use *Stuff That Works!* usually require students to record their activities and reflections in journals.

For many of the *Stuff that Works!* units, the opening activity is a scavenger hunt or brainstorming session. In a brainstorming session, students think of the examples, list them, and then try to make sense of them. Often, the teacher starts the discussion by asking students to tell what they know about the meaning of a word. These discussions can be rich opportunities to explore and inquire about language. For example, at the beginning of their units on packaging, both Verona

Williams and Roslyn Odinga asked their students, "What is a package?" In Roslyn's second grade class, this question led to an extended discussion about how and whether the outsides of various fruits were examples of packages. The students in this class were "draw[ing] on their prior experience, their interactions with other readers and writers [and] their knowledge of word meaning and of other texts," to make connections between this new word and others that they already knew. (ELA Standard #3, p. 31)

Packaging and Other Structures also engages students in recognizing storage problems in their classroom, and in designing and testing ways to store things better. These design projects require considerable discussion, as well as more formal oral, written, and graphic presentations. First, students identify the problem they want to solve, either through a brainstorming session or because it is already an obvious concern. Next they decide on the kinds of information they need to understand the problem better. At some point, they brainstorm about the criteria that a successful design would have to meet; in other words, what the design would have to do in order to solve the problem. Subsequently, they meet in small groups to come up with possible solutions. The groups then implement their designs, and eval- uate how well they meet the criteria.

Each of these steps engages students in using "spoken, written, [or] visual language to accomplish their own purposes." (ELA Standard #12, p. 45) The purposes are genuinely the students' own, because the design projects address problems they have raised. To accomplish their goals, they have to brainstorm about the criteria the design should meet, how to collect data, and how to test the design. They have to negotiate with one another to come up with a solution everyone can accept. They have to present their ideas to one another in written, oral and graphic forms. The design of useful classroom structures provides rich opportunities for developing language proficiency.

APPENDIX A

Packaging Unraveled: The Inside Story

This section answers some of the questions raised in Chapters 1 and 2 and provides additional background material about packages and packaging materials.

Cardboard Heroes

Cardboard is so widely used in packaging that the word *package* is often assumed to mean "cardboard package." It is estimated that about 90% of all manufactured goods in this country are at some point packaged in cardboard. Packaging includes more than the container that you see in the store, put in your shopping cart, and bring home.

Most prepackaged goods arrive at the supermarket on large wooden platforms called skids or pallets. On a skid, there are typically several dozen cardboard boxes held together with metal straps and/or shrink-wrap plastic. A typical carton may hold half a dozen to two dozen individual jars, bottles, or smaller cartons, which are the packages the consumer sees. On their journey from the manufacturer to the store, most consumer products travel in cardboard.

What, exactly, is cardboard? To a paper manufacturer, "paper board"— the industry term for cardboard—is nothing but a heavier, coarser grade of paper. Some cardboard comes in flat sheets, like the material of index cards, but most cardboard is sold in *corrugated* form. Corrugated cardboard is built around a fluted layer, also called the medium. It is the only part that is actually corrugated.

A corrugated layer or medium is shown in Figure A-1(A). The medium is then glued to a flat sheet, which the industry calls a facing, as shown in Figure A-1(B). This material, which is called single-face corrugated, is never used to make a carton, but is sometimes used as cushioning material— for example, to wrap glasses or mugs. Usually, another liner is added to the other side, as in Figure A-1(C), to make double-faced, or, as it is more commonly known, singlewall corrugated. This is the material used to make about 90% of all cardboard cartons. For heavier cartons, another medium and facing are added to make "doublewall corrugated." Figure A-1(D) shows a piece of doublewall cardboard. As we have seen, the purpose of using corrugations is to make a structure that is strong as both a column and a beam (see Figure 2-20).

Most cartons have a box certificate printed on them, which provides information about the cardboard they are made of. Figures A-2 and A-3 show the certificates from two singlewall cartons. In both certificates, the last two lines give the "Size Limit" and "Gross Weight Limit," respectively. These limits are set for a particular grade of cardboard by truck and rail freight regulations. The size limit refers to the sum of all three dimensions of a rectangular box: the height + width + depth cannot exceed this limit. The weight limit includes the box itself, as well as the contents.

The top line of the box certificate comes in two varieties. Figure A-2 shows the most common type, whose first two lines indicate "Bursting Test" and "Min. Comb. Wt. Facings (Minimum Combined Weight Facings)," respectively. The Bursting Test gives a measure of how well the carton can stand up to pressure on its side. The pressure in a fluid is increased gradually until a rubber diaphragm bursts through the cardboard. The pressure needed to burst through, in

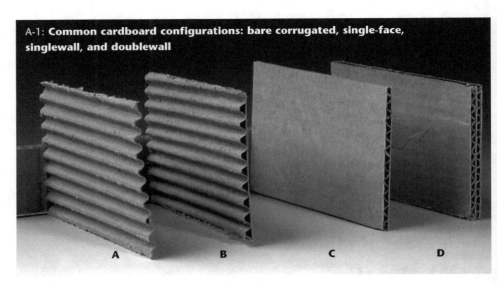

A-1: **Common cardboard configurations: bare corrugated, single-face, singlewall, and doublewall**

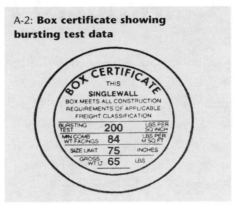

A-2: **Box certificate showing bursting test data**

A-3: **Box certificate showing edge crush test data**

pounds per square inch, is recorded as the "Bursting Test" data. It is a test of how strong the side, top or bottom of the box is when it is loaded.

The second line, "Minimum Combined Weight Facings," records the weight (in pounds per thousand square feet) of the two facings (flat sheets) of this singlewall cardboard. For this particular material, the sum is 84, or 42 lbs/1000 sq. ft. per facing (Figure A-2). By comparison, ordinary copy or printer paper weighs about 15 lbs./ 1000 sq. ft, or about one-third as heavy as the cardboard liner.

The box certificate in Figure A-3 is different. Instead of the Bursting

Test and Minimum Combined Weight Facings, it has a single line called "Edge Crush Test (ECT)." This intriguing process consists of mounting a two-inch wide by inch-and-a-quarter high strip of the material so it is standing up, and pressing it with a machine until it buckles. The weight that it takes to crush the sample, divided by the width of its edge, is the ECT result in lbs./in. The ECT actually gives a better measure of the strength than the bursting test, because boxes most often fail by buckling at the bottom of a heavy stack. The ECT is a measure of how well the side of the box works as a column.

By far the most common type of cardboard container is the folding carton, which is delivered flat to the product manufacturer, to save shipping space. They are nearly always made of corrugated cardboard. Currently, the cardboard for these cartons is most likely cut and scored by automatic machinery, which helps to explain the relatively small number of standard shapes and sizes.

The form of most folding cartons is so common that the industry calls it the "Regular Slotted Container (RSC)." An example is shown in Figure A-4. The style of this box is sometimes called "the tube" because the sides are joined together, but the end flaps are initially open. The side joints may be made using glue, tape, metal staples, or metal stitches. In Chapter 2 we explored some of the pros and cons of these joining methods. To see why the RSC is so common, look at the folded out version in Figure A-5. It is almost perfectly rectangular. If any of the flaps were longer or shorter, more complicated dies would be required to make these cartons, and more cardboard would have to be thrown away to make the flaps.

A common alternative to the Regular Slotted Container is the same box with flaps on one side only. This type of box, which also uses tube construction, is called the "Half Slotted Container (HSC)." Because it is lacking one set of flaps, another piece is needed to form the lid, which has to be a little larger than the bottom. The lid could be another HSC, or another type of construction called the tray, which has a solid bottom or top and folding sides. Figure A-6 shows a Half-Slotted Container with a tray lid.

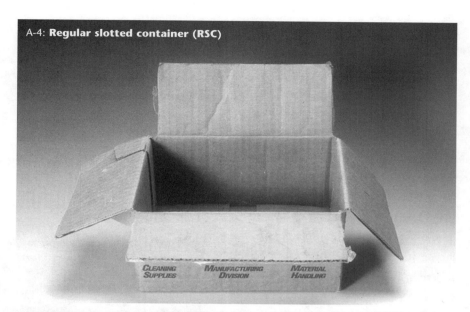

A-4: **Regular slotted container (RSC)**

A-5: **The same carton, folded flat**

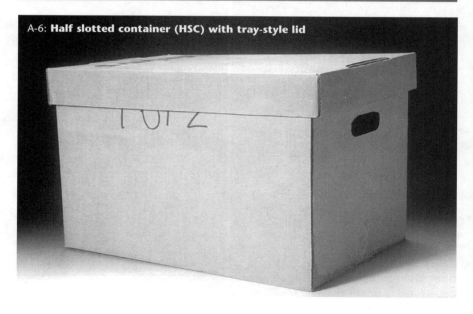

A-6: **Half slotted container (HSC) with tray-style lid**

The cardboard boxes described so far all require tape, glue, or metal fasteners to assemble them. There is a whole category of boxes that can be folded flat and assembled completely without using any of these joining methods. These boxes depend completely on clever design and friction to hold them together. Examples are shown in Chapter 1 and 2 in Figures 1-2, 1-18, 1-19, and 2-34. Note that both the bottom and the lid are of "tray" construction, and that they are hinged. There are all sorts of folding box designs, which fold and unfold without any joining, including tubes, trays, unusual shapes, cutouts. You can make your own collection of assorted folding boxes and examine how each one works.

In addition to the folding carton, there is a more expensive kind of box called the "rigid setup box." These are assembled by the box manufacturer and delivered to the product manufacturer ready to be filled (Figure A-7). They usually have a nice decorative finish to them. They are usually of the tray style of construction, with the sides attached by tape along their edges. Some setup boxes have hinged lids, while others have separate tops and bottoms. Cigar boxes are the most famous type of setup boxes with hinged lids.

A-7: **Rigid setup box with separate lid**

How Shopping Bags Fail

Shopping bags can and do fail unexpectedly, sometimes on the way home from the store. In Chapter 2, we outlined some suggestions for testing the strength of shopping bags. In this section, we will discuss three of the most common failure modes of shopping bags, and show how the concepts of tension, compression, and shear are useful in understanding them.

The most common type of plastic shopping bag has handles that are simply cut out of the body of the bag. The top may be cut square with notches for the handles below, or the handles may be raised above the top, as is the case with most supermarket bags. In either case, the most common type of failure, as every shopper knows, is for one of the handles to stretch and eventually tear apart when the bag is too heavily loaded (Figure A-8).

The tearing of the handles is an example of failure in tension. It occurs because of forces that are pulling away from one another in opposite directions, as shown in Figure A-11(A). The force pulling upward comes from the hand holding the bag. The force pulling downward results from the weight inside. The handles are stretched because they are the narrowest part of the bag, and all of the force from the weight is concentrated there. As they become longer, they also become thinner, just like a rubber band does when stretched. As the material becomes thinner, the force becomes even more concentrated, until the material simply isn't strong enough to manage so much force over such a small surface. At that point, the handles literally pull apart.

Nearly all paper shopping bags, and some plastic shopping bags, have handles that are made separately and then attached to the body of the bag. In plastic bags, these handles are usually "heat sealed" to the body—i.e., attached by briefly melting a small section of both body and handle together. With paper bags, the handles are nearly always glued. To make the glue joints stronger, paper patches are often glued to both the handle and the body of the bag.

A-8: **Shopping bag post-mortem: tension failure of handles**

Figure A-11(B) shows how the glued handles usually come off paper bags. This is a typical attachment, which has to support shear. When the shear load becomes too great, the glue joint between the patch and the handles fails, and the handle simply pulls out. An example is shown in Figure A-9.

Nearly all paper bags have glued joints at the bottom, as well as at the handles. A common type of paper bag failure happens when the glue on the bottom stops working. The chain of events leading to bottom failure is more complex than the others, involving both tension and shear. An example is shown in Figure A-10.

As the end view in Figure A-11(C) shows, the upward force on the handles opposes the weight of the object in the bag. The bottom of the bag then acts like a beam. Its bottom surface tends to spread outwards, in tension, like the beam in Figure 2-18. The pulling outward is shown in the bottom view in Figure A-11(C). These opposing forces eventually make the bottom fail. One way for it to fail is for the glue joint to come apart—an example of shear failure. However, the glue may be stronger than the paper itself. Iin this case, the paper wil tear first, which is an example of tension failure. In Figure A-10, you can see places where the glue joint failed, as well as a few points where the paper tore.

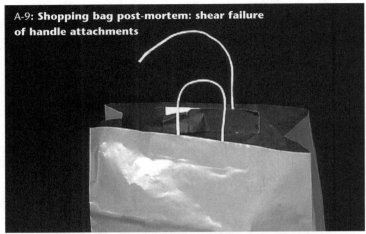

A-9: **Shopping bag post-mortem: shear failure of handle attachments**

A-10: **Shopping bag post-mortem: complex failure of bottom**

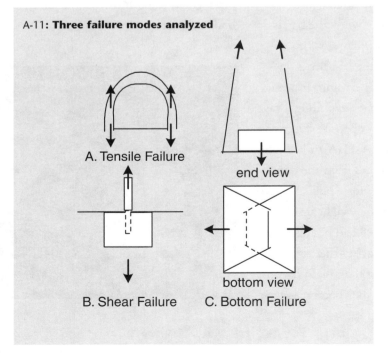

A-11: **Three failure modes analyzed**

A. Tensile Failure

B. Shear Failure

end view

bottom view

C. Bottom Failure

Anatomy of a Can

Until around 1970, nearly every metal can was made of steel. Although popularly known as "tin cans," there is only a very thin layer of tin plating, to protect the steel from rusting. They are also lacquered, for further protection. These cans are made in three pieces: the body, the top, and the bottom. The body is made by wrapping a rectangular piece of metal into a cylinder, and then joining the two ends. The top and the bottom are then attached separately. The side seam was sealed with a solder made mostly of lead. Recently, concern over lead poisoning has led to strict regulation of lead in food and beverage packaging, and the phasing out of solder-sealed cans. Nearly all three-piece cans now use welded side-seams.

Aluminum is not nearly as strong as steel, but it does have some major advantages as a can material. Because it is softer, a disk of aluminum can be formed into a cup-shaped bottom by punching a deep depression in it. The only remaining piece is the top, so these are referred to as two-piece cans (Figure A-12). The top can be scored, allowing it to be opened with a pull-ring, which is not possible for steel tops. Finally, aluminum cans weigh about half as much as comparably sized steel cans, which lowers shipping costs.

A-12: **Stages in making a two-piece can**

A-13: **Aluminum can showing domed bottom**

The walls of an aluminum soft-drink can are extremely thin, as you can tell by seeing how easy it is to crush an empty can. (Crushing steel cans used to be a trick for weight lifters only!) The reason that the walls can be so thin is that the internal pressure of the soda helps to make the can much more rigid. However, this pressure could also make the bottom of the can bulge out. If this were to happen, the can would not have a flat bottom to stand on. (A similar problem exists for plastic soda bottles, as we shall see in the next section.) To counteract this tendency, the bottom of the can is domed, as shown in Figure A-13. As we saw in Chapter 2, the dome is a very strong structure, which resists the compressive forces coming down from cans stacked above.

Another interesting feature of the aluminum beverage can is the top. Why is it so much smaller in diameter than the body of the can (Figure A-14) ? The answer lies in the relative cost of the top, compared with the rest of the can. In order for the pull-tab to open the top, a small section of the top needs to be scored, so that it will break away easily. However, there has to be enough material under the score lines so that the can does not open prematurely. For the top to accommodate just the right thickness of scoring, it has to be considerably thicker, and therefore more expensive per square inch, than the rest of the can. Necking the top allows the can manufacturer to get by with less of this more expensive material. How much does this really save? For one can, not much, but what if you multiply by the 100 billion cans produced annually in the U.S.?

A-14: Aluminum can showing necked top

Decoding Plastic

Until the 1960s, very little packaging consisted of plastic. Since then, however, plastic has taken over an increasingly large share of the packaging market. More than a third of all plastic is used for packaging. It has some major advantages over cardboard, metal, and glass: it is cheap, easy to form into virtually any shape, and much less breakable than glass. Unlike cardboard, it is waterproof, and generally resistant to chemicals. Unlike metal, it does not corrode, and it is lighter than either metal or glass. Plastic can be colored or transparent, and most plastic used for packaging is also recyclable.

To aid in recycling, most plastic containers produced in the U.S. are stamped with recycling numbers. An example is shown in Figure A-15. What do these numbers mean? What information can be learned from looking at these numbers and the containers they appear on?

Recycling number 1 indicates Polyethylene Terephthalate, PET or PETE for short. PET is the only plastic material used to make carbonated beverage bottles. In 1977, Pepsi Cola invented a process for making PET bottles strong enough to hold soda. It involves blowing air into a heated slug of plastic, until it expands against a metal mold.

A-15: **Recycling number on the bottom of a plastic container**

A-16: **Bottom view of plastic soda bottle**

PET is also used to make audio and video tape, decorative balloons with metal coatings, and "fleece" vests and pullovers.

Recycling number 2 represents High-Density Polyethylene (HDPE). While PET is strong, the manufacturing process makes it impossible to add jug-style handles to the container. Containers with handles, such as one-gallon milk containers, are therefore made of HDPE. This material is also very resistant to strong chemicals, so it is also the material of choice for detergents and other cleaning fluids. However, HDPE is not transparent, so it cannot be used to hold any product that has to be visible through the container.

Recycling number 3 stands for Polyvinyl Chloride, or PVC, commonly known simply as "vinyl." At one time it was used for making liquor bottles, and was also a candidate for soda bottles. Currently, this material is not generally allowed for making food and beverage containers because it is suspected of causing cancer. However, some one-gallon water bottles are now made of PVC, which is the clearest of all common plastics. You can find PVC in many containers for hand soaps, shampoos, and other beauty care products.

Recycling number 4 symbolizes Low-Density Polyethylene or LDPE, which is chemically similar to HDPE (#2), but is much softer and more flexible. It frequently used for tops,

One problem with the method is that the walls become fairly thin. As a result, the bottom of the bottle has a tendency to bulge out, which can prevent it from standing up straight. The same problem was discussed above for aluminum soda cans. Until the early 1990s, a separate base of blue or green plastic was attached to the bottom of soda bottles to make them stand up reliably, but this required several extra steps in manufacturing. More recently, a process was developed that adds plastic to the base while the bottle is being blown. Additional plastic is injected at the very bottom of the base, which makes the base thicker, and forms the little feet which you can find on the bottom of any soda bottle (Figure A-16). The little button of plastic at the center (see arrow) is where the extra plastic was injected in.

One of the ironies of recycling numbers is that the "1" rarely appears any more on the bottom of soda bottles. The reason is that all soda bottles are made of this stuff, so recyclers already know what it is!

such as caps of plastic soda bottles. Because LDPE melts at a much lower temperature than PET, the top can be heated enough to be formed around the bottle, without softening the bottle. The pint- and quart-sized tubs for take-out food usually have tops of LDPE and bases of PP (#5 below). The bottoms can be microwaved, but the tops cannot! LDPE is also used to make garbage bags, most plastic shopping bags, and bubble wrap.

Number 5 indicates that the product is made of Polypropylene (PP), which is usually found in a semi-rigid, semi-transparent form. As mentioned above, PP is used to make take-out tubs, and also smaller tubs, such as those used for margarine and yogurt. Six-pack ring-holders are made of this stuff, as are cheese wrappers, inside liners for cereal and cracker boxes, medicine bottles, toys and a host of other products. PP can be microwaved.

Number 6 indicates Polystyrene (PS), which is the cheapest plastic. This is the material used to make those flimsy cookie trays, salad bar containers, thin transparent cups, and virtually anything else made of very thin plastic. Polystyrene also comes with the trade name Styrofoam, which is the same stuff with little gas bubbles trapped inside. Foams like Styrofoam are good for cushioning, because the little gas pockets allow the material to change shape and absorb impacts (see next section). The gas pockets also make Styrofoam an excellent thermal insulator. For this reason, Styrofoam is used to make insulated "hot" cups, insulated food containers, and coolers.

What Happens in a Crash

Why do some objects need to be cushioned, and others don't? What kind of material is good for cushioning and why? Fashion a ball of clay into a cup shape and drop it on the floor. It will change shape a little, but will probably stay intact. If the same piece of clay has been fired in a kiln, to make a pottery cup, it will probably break if you drop it on the floor. Silly Putty feels like clay, but if you drop it on the floor, it will probably bounce. Why do some materials change shape on impact, while others break and still others bounce back? Thrown hard, a rubber ball will bounce off the floor, get buried in the mud, or shatter a glass window. Why does the same rubber ball have such a different effect on these three materials?

The answers to these questions go straight to the heart of the atomic structure. Materials that bounce back, like rubber, are said to be *elastic*. These materials have long coiled chains of atoms. When compressed or stretched, these chains become more or less tightly coiled, but when released, the coils return to their original shapes. Materials that break easily like glass, chalk or baked cookies are known as *brittle* materials. In these materials, microscopic cracks create zones of weakness between layers of atoms, which relatively small loads can open up like a zipper, resulting in almost instant fracture. Soft, pliable materials such as clay, wax, or dough are considered *ductile*. When they are subjected to loads, layers of atoms shift and slip against one another, leading to a permanent change in shape rather than immediate fracture. Materials behave very differently because of differences in the ways their atoms and molecules are organized.

Suppose a box with an object in it is dropped. The object is likely to break if it is brittle, bounce back if it is elastic and change shape permanently if it is ductile. A common word for brittle is "fragile." When a fragile object is transported, there is always the possibility that someone might drop the package, or that there might be an impact from the side or top. Any of these impacts could break the object. The purpose of cushioning is to prevent it from breaking.

Cushioning materials work in a number of different ways to protect fragile objects from impacts. A coffee cup has a handle that is much weaker than the rest of the cup. Cushioning material helps to insure that the cup never rests on its handle, and that any impact absorbed by the handle is spread to the rest of the cup as well. By filling up all of the space inside a package, cushioning also prevents a fragile object from moving around and crashing into another object or the wall of the carton. Many cushioning materials have little air spaces inside them. These include corrugated cardboard, Styrofoam, foam rubber, and straw (Figure A-17). These materials not only fill up space but can also absorb some of the energy of an impact themselves by tightening up their own air pockets (Figure A-18).

For products that are not so fragile and that are fairly light, space filling may be the only requirement. Space fillers include particles such as sawdust or Vermiculite, shredded paper, crepe paper, and bubble wrap. These materials mostly deform permanently on impact. For example, in a serious impact, the bubbles in bubble wrap may pop. As a result, they cannot be relied upon to absorb much energy. Space-filling materials are used for small, relatively sturdy products. For example, the inside of a book-mailing envelope is filled with shredded paper or bubble wrap.

The most fragile objects require elastic cushioning that will absorb considerable amounts of energy, and do so repeatedly, in case the package is dropped more than once. The best materials for this purpose are foam rubbers, which are made of plastics such as polyurethane. When subjected to an impact, the cavities in the foam change shape, as shown in Figure A-18. Afterwards, they return gradually to their original shape. Because the original shape is restored, elastic materials are also ready for the future impacts. Light, very fragile objects, such as small electronic products, watches, and jewelry, are likely to be shipped in foam rubber.

A-17: **Box with straw cushioning**

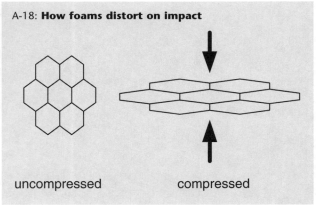

A-18: **How foams distort on impact**

uncompressed   compressed

Elastic materials are ineffective if the product is too heavy. A heavy product will pre-compress foam rubber just by resting on it, so that it can't compress any further on impact. For heavy objects, a semi-elastic cushioning material such as Styrofoam is needed (Figure A-19). Styrofoam is rigid enough to support a considerable amount of weight without compressing. It can then absorb moderate impacts and return part of the way to its original shape. Styrofoam can't absorb as much energy as foam rubber, but it doesn't need to, if it is being used to cushion something heavy. Because heavy objects can't easily be lifted very high, they aren't normally dropped as far. Styrofoam blocks, slabs, or molded caps are used to cushion large electronic products such as TV sets and computers.

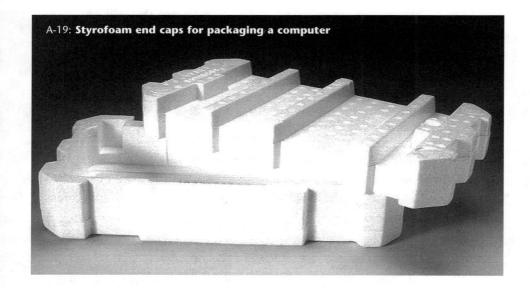

A-19: **Styrofoam end caps for packaging a computer**

How Pump Dispensers Work

A-20A

A-20B

Pump dispensers are sold as part of the packaging of many thick liquids, ranging from hand lotion to mustard. Most people throw them out with the empty bottle, but these actually are fascinating devices that are worth exploring. A way to start is to keep careful track of what actually happens when you operate one of these pumps.

Figures A-20 A through F show the sequence of events in detail. In A, an empty pump dispenser is placed in the fluid for the first time. The plunger is pushed down in B, and then released in C. Notice that as the white plunger is released, the chamber just below it half fills with the fluid. Still, nothing comes out of the spout. Next, the plunger is depressed again, in panel D,

and released in E. This time, the chamber becomes entirely filled with the fluid. Then the plunger is pushed down again in F, and fluid comes out of the spout!

There are really two separate operations going on here. First, the chamber has to be filled with fluid. This operation only happens on the upstroke. In this particular pump dispenser, it takes two upstrokes to fill the chamber completely. Other pump dispensers may require more or fewer strokes to accomplish the same thing. Once the chamber is full, the device is ready for the second operation: forcing the fluid out of the spout. When the plunger is pushed into a full chamber, there is no place for the fluid to go except out through the spout.

A-20C

A-20D

A-20E

A-20F

How does the pump dispenser do all of this? To answer this question, we'll need to look at two important details of the way it is constructed:

1. **The plunger is hollow** (Figure A-22A). This allows fluid to pass through it when it is pushed down.

2. **A little ball bearing is held in a little space between the chamber and the tube that goes down into the tank of fluid** (Figure A-21). This little ball forms a one-way valve. which prevents air or liquid from ever passing downward from the chamber into the tank. Liquid can only pass upwards through the ball valve.

Based on this information, we can now explain all of the steps illustrated in Figure A-20. The six diagrams in Figure A-22, A through F, correspond to the photos in Figure A-20, A through F, respectively.

When you first push the plunger down (Figure A-22B), the ball valve closes, preventing air from passing into the tank. The valve closes because gravity keeps the ball down, closing off the opening into the tube. As a result, you will not notice any air bubbles in the tank, as you would if you pushed

air into the fluid by blowing into a straw or squeezing an eyedropper. When the plunger is released, in C, the ball valve opens, because now the suction can lift the ball out of its seat. Therefore, the chamber partially fills with fluid. A similar sequence occurs during the second down stroke and up stroke, in D and E, respectively. At this point, the chamber has completely filled with fluid. The next down stroke, shown in F, forces the fluid through the hollow plunger and out the spout. It can't go back down into the tank, because the bali valve blocks it, so it can only go up, and it finally squirts out.

A-21: Detail of pump dispenser, showing ball valve

A-22: How a pump dispenser works

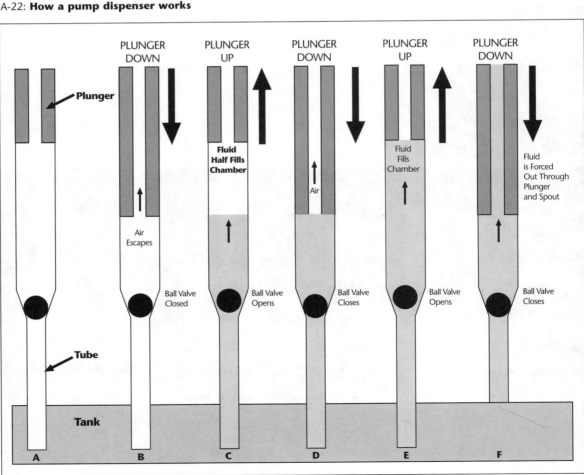

REFERENCES

Chapter 2 and Appendix A

Barker, Marilyn, ed. (1986) *The Wiley Encyclopedia of Packaging Technology.* New York: J. Wiley & Sons.

This authoritative source has considerable material that is accessible to the non-technical reader. The articles on "Boxes, Corrugated," "Cartons, Folding," and "Closures, Bottle and Jar" are particularly useful.

Cassidy, John. (1985) *Klutz Book of Knots.* Palo Alto, CA: Klutz Press.

Knots are often the best way to join larger structural elements such as bamboo and rolled-up newspaper. Nearly anyone can learn to tie the most useful and common knots using this step-by-step, multicolor, humorously written guide. It comes with string and notches for practicing.

Fenichell, Steven. (1996) *Plastic: The Making of a Synthetic Century.* New York: HarperCollins Publishers.

Here is a popular account of the invention, uses, and consequences of the major plastic materials.

Gordon, J.E. (1986) *Structures, or Why Things Don't Fall Down.* New York: Da Capo Press.

This book offers an entertaining and often humorous account of how and why artificial and natural structures are organized the way they are. Gordon discusses how bats, trees, planes, ships, and dresses operate as structures.

Hanlon, Joseph F. (1984) *Handbook of Package Engineering* (2nd Edition). New York: McGraw-Hill Book Co.

This small book offers a wealth of information, much of it non-technical, about packaging materials and manufacturing technologies.

Henessey, James and Papanek, Victor. (1975) *Nomadic Furniture 1.* New York: Pantheon Books.

This book shows you how to make temporary or easily movable furniture, some of it from recycled materials.

Hine, Thomas. (1995) *The Total Package: The Evolution and Secret Meanings of Boxes, Bottles, Cans and Tubes.* Boston: Little, Brown and Co.

Hine provides a history of modern packaging, and shows how it has evolved together with changes in society. It includes beautiful photos of packages from an earlier era. The focus is primarily on the promotional aspects of packaging.

Kluger-Bell, Barry. (1995) *The Exploratorium Guide to Scale and Structure: Activities for the Elementary Classroom.*
Portsmouth, NH: Heinemann.

This is a carefully thought-out, well-sequenced curriculum on structures for the upper elementary grades. It includes a brief but very accessible section on science and engineering principles, a wealth of activities, teacher tips and connections with standards. All of the activities use common materials.

Mumford, Lewis. (1967) *Technics and Human Development: The Myth of the Machine, Volume 1.*
New York: Harcourt Brace Jovanovich.

This classic interpretation of the origins of technology emphasizes the role of language and art, far more than most accounts. It also criticizes the male bias in most histories of technology.

Package Design in Japan. (1993) Cologne, Germany: Taschen.

This gorgeous book shows examples of how the Japanese think about packaging. Every package in this book is a work of art, intended to have at least as much value as the contents.

Paine, F. A., ed. (1991) *The Packaging User's Handbook.* New York: Van Nostrand Reinhold.

Here is a fairly non-technical compendium of information about all the major problems in packaging. The chapters on "Closures and Dispensing Devices" and "Package Cushioning Systems" are particularly useful.

Petroski, Henry. (1992) *To Engineer Is Human: The Role of Failure in Successful Design.* New York: Vintage Books.

Petroski uses examples from the history of structural engineering to show how design progresses via analysis of past failures. The author suggests experiments with everyday objects like kitchen knives and paper clips.

Petroski, Henry. (1996) *Invention by Design.* Cambridge, MA: Harvard University Press.

Chapter 5, "Aluminum Cans and Failure" provides a brief history of beverage cans, and dicusses the invention of the pop-top in detail.

Roth, Laszlo and Wybenga, George L. (1991) *The Packaging Designer's Book of Patterns.*
New York: Van Nostrand Reinhold.

This book contains hundreds of drawings, showing the flattened and assembled views of a wide variety of folding boxes. It also contains some technical information on cardboard.

Salvadori, Mario and Tempel, Michael. (1983) *Architecture and Engineering: An Illustrated Teacher's Manual on Why Buildings Stand Up.* New York: New York Academy of Sciences.

Some 70 lessons, aimed at junior high school and high school levels, describe step-by-step how to build and test structures from everyday materials.

Salvadori, Mario. (1990) *Why Buildings Stand Up: The Strength of Architecture.* New York: W. W. Norton.

This classic is a very readable account of how structures support loads, and why they sometimes fail to. It provides clear explanations of compression, tension and shear, and describes major structural elements, such as arches, columns and beams.

Schools Council 5/13 Series. (1982) *Structures and Forces: Stages 1 and 2.* London: MacDonald Educational.

This excellent volume is full of ideas about how children can observe and experiment with structures. There are excellent examples of investigations designed by children, as well as ideas for making and testing structures using common materials.

Vogel, Steven. (1998) *Cats' Paws and Catapults.* New York: Norton.

Vogel, a biologist, has written a clear, comprehensive account of the major differences between the mechanical worlds of nature and technology. Chapters 4 through 7 deal with structures.

Wilson, Forrest. (1989) *What It Feels Like to Be a Building.* Washington: Preservation Press.

This unique book has very few words, but many, many pictures, and is suitable for all levels. There are drawings of famous structures next to silhouettes of people supporting each other in the same way as these structures.

Zubrowski, Bernie. (1993) *Structures.* White Plains, NY: Cuisenaire Co.

This compact guide provides instructions for building and testing a variety of straw structures: a house, a column, a bridge and a tower. Although the activities are intended for grades 5-8, the book is an excellent resource for all grade levels.

Chapter 6

American Association for the Advancement of Science. (1989) *Science for All Americans: A Project 2061 Report on Literacy Goals in Science, Mathematics and Technology.* Washington, DC: Author.

American Association for the Advancement of Science. (1993) *Benchmarks for Science Literacy.* New York: Oxford University Press.

American Association for the Advancement of Science. (1997) *Resources for Science Literacy.* New York: Oxford University Press.

American Association for the Advancement of Science. (1998) *Blueprints for Reform.* New York: Oxford University Press.

American Association for the Advancement of Science. (2001) *Designs for Science Literacy.* New York: Oxford University Press.

International Technology Education Association. (1996) *Technology for All Americans: A Rationale and Structure or the Study of Technology.* Reston, VA: Author.

International Technology Education Association. (2000) *Standards for Technological Literacy: Content for the Study of Technology.* Reston, VA: Author.

National Center on Education and the Economy. (1997) *New Standards Performance Standards; Vol 1: Elementary School.* Washington, DC: Author.

National Council of Teachers of English & International Reading Association. (1996) *Standards for the English Language Arts.* Urbana, IL: Author.

National Council of Teachers of Mathematics. (1989) *Curriculum and Evaluation Standards for School Mathematics.* Reston, VA: Author.

National Council of Teachers of Mathematics. (2000) *Principles and Standards for School Mathematics.* Reston, VA: Author.

National Research Council. (1996) *National Science Education Standards.* Washington, DC: National Academy Press.

Arch: A rounded structure that supports loads largely in compression.

Beam: A horizontal building element.

Box certificate: A standard label printed on a cardboard carton with information for shippers and freight handlers about the characteristics of the cardboard.

Buckling: Failure of a column by bending near the middle.

Column: A vertical building element, designed to resist compressive loads.

Compression: The state of a material made more compact by pressure.

Compression structure: A structure that supports loads by resisting compression.

Corrugated: Having a wavy shape; also, a name for cardboard that has a wavy layer in between two flat layers.

Dead load: The weight of a structure itself, not including the load it is intended to support (see *live load*).

Dome: A cylindrical structure, rounded at the top, which is the equivalent of having an arch in every direction.

Equilibrium: A balance of forces, which keeps an element from moving.

Facing: One of the two flat sides of a piece of corrugated cardboard.

Failure: Situation that occurs when a structure can no longer resist the loads it is subjected to.

Forces: Pushes or pulls on objects.

I-beam: A structural member, whose cross-section looks like the capital letter "I."

Inclined plane: A ramp used to lessen the amount of force needed by increasing the distance over which a load must travel.

Live load: The weight of anything supported by a structure.

Load: A force, such as gravity, that a structure must withstand in order not to fail.

Mechanism: A device with moving parts that converts force and motion at one point to a different combination of force and motion at another point.

Medium: The corrugated middle layer of a piece of corrugated cardboard.

Packaging: Material used to contain, protect, and make it easier to transport an item.

Recycling number: A number from one to seven that appears on the bottom of a plastic container.

Shear: State of a material that is acted upon by off-center forces; for example, the forces on a deck of cards when you press your hand across the top of the deck.

Shear resistance: Ability of a material to resist shear forces.

Simple machine: One of several elementary machines once considered to be the elements of which all machines are composed: the lever, the wheel-and-axle, the pulley, the inclined plane, the wedge, and the screw.

Splaying: Failure of columns to remain vertical by spreading or slipping outward.

Stability: The capacity of an object to return to its original position after having been displaced.

Strength: Resistance to failure.

Structure: A device or system designed to withstand a load.

System: The arrangement or interrelation of all of the parts of a whole.

Strut: A structural element, such as a column, intended to resist compression.

Technology: The artifacts, systems, and environments designed by people to improve their lives.

Tension: The state of a material produced by the pull of forces, tending to cause it to extend.

Tension resistance: Ability of a material to resist being extended, without breaking or stretching irreversibly.

Tie: A structural element designed to resist tension, such as a fishing line.

Viscosity: The property of a fluid that describes the difficulty encountered in making it flow.